P9-AQN-826

pointed the way to likely mineral deposits, oil fields and other natural resources. They have shown the extent of crop disease, forest fires and damage from erosion. They have even advanced the cause of peace, by making on the spot inspection of missile sites unnecessary. Spy satellites are providing intelligence essential to our national security.

The earth is still unknown and unpredictable; but the broader range of vision and data offered by satellites has greatly increased man's chances for survival and control of his environment. The space program may ultimately make the difference in solving worldwide problems like pollution and finding new sources for energy and food.

about the author

Günter Paul is a physicist at the Institute for Extraterrestrial Research of the University of Bonn. Born in West Germany, he is a member of the German Society for Air and Space Flight, and gives regular popular lectures on space exploration.

Translated by Alan and Barbara Lacy

Günter Paul

The Satellite Spin-Off

The Achievements of Space Flight

Robert B. Luce Washington — New York

Paul, Günter, 1946-
The satellite spin-off.

Translations of Die dritte Entdeckung der Erde.
Includes index.
1. Scientific satellites. 2. Artificial satellites.
3. Remote sensing systems. I. Title.

TL798.S3P3813 629.43′4 75-11369
ISBN 0–88331–076–7

Contents

Foreword

Of what use is the exploration of outer space? Who has not some time or other asked himself this question at least once? According to surveys, people feel that exploration of outer space is promoted much too much, while ignoring man's problems. Was the German Nobel Prize winner, Max Born, right when he said, "I do not see that exploration of space contributes anything to the material well-being of man?"

This author is of a different opinion. Satellites such as Echo, Telstar, Tiros, Early Bird, and recently ERTS have ushered in a new age. Our planet is shrinking. At the time of the great explorations, and even much later, contact with distant lands was an adventure reserved for a select few. When the airplane, the telegraph and the telephone were invented, a rapid change took place. Suddenly America was no longer quite so far from Europe. New territories opened up.

Now we stand on the brink of the discovery of the world for the third time. Europe lies at our front door, no further away than Canada or Mexico. A telephone call can be placed directly from New York to Bonn without operator assistance. Weather satellites daily provide a global survey of the earth, making us lose sight of the distances between countries. New rivers and lakes are discovered from outer space, crops are kept under observation, and areas which might be expected to contain rich sources of raw materials are surveyed, no matter how remote and inaccessible.

We are gradually realizing that we are all citizens of one

planet. Space flight has given us this insight. It is no longer the moon that is the focus of interest; it is our earth. We all profit from space flight, directly or indirectly. Can we forget that July 23, 1962 when 200 million viewers first experienced a "cosmic" television broadcast? The significance of the Atlantic was thereby reduced to that of an inland sea.

July 23, 1962 did not remain unique. In 1967 alone, satellites of the International Telecommunications Satellite Consortium, INTELSAT, transmitted 200 hours of live television broadcasts. The world was shrinking. We were there when a new president was elected in the United States, an airplane crashed in Japan, or when the Olympics were taking place in Munich. This book is about the rediscovery of the world, about the benefits of space flight.

The well known by-products of space flight will not be considered. We will emphasize the successes of communications, weather, navigational and geological satellites. We are concerned with treasure hunts from outer space, supervision of harvest yields, the opening up of inaccessible regions. Unfortunately, many possibilities are still unexplored. Consequently the points of emphasis in this book are historically determined, although many readers would probably like to see greater stress on the geological aspects. Let us not forget, however, that this phase of space travel stands at the beginning of a fascinating development whose dimensions cannot even be estimated.

The author is aware that he would not have done justice to the subject if he had attempted a complete history of the so-called practical satellites. Concrete examples, stories from the daily lives of the satellites, illustrate their usefulness far better. For this reason, the stage for this book is the whole world: from the eskimos of Canada to the farmers of India, from Japan to Peru.

It is a well-known secret that the military also profit from space exploration. They have their own satellites, and could no longer exist without them. There is a great deal of speculation about these satellites. What are the military observers able to see on earth from space? This kind of book requires that these questions also be explored. After all, they concern the direct application of satellites. The possibility of preventing wars through this type of reconnaissance cannot be completely discounted. At any rate, it would be an achievement if the exploration of outer space were able to contribute in this respect towards solving world problems.

1. Reflectors in the Sky

A Brief Prologue

The night of February 16, 1964 is cool, a typical winter night. The clouds are gathered in a few places in the sky, and the weather is suitable for observation. Seldom do the stars twinkle so clearly as on this night. Yet it is not for this reason that several people have ventured out into the streets. No, they want to see a rare spectacle, the meeting of two satellites — Echo 1 and Echo 2. The teletype message from the Bochum observatory stated it quite clearly: "During the early morning hours there will be an opportunity to observe a remarkable sight. Glowing brightly, two satellites will speed across the sky in different directions a few minutes apart. On February 17 after 2:00 A.M. Central European Time it is expected that both satellites will be visible simultaneously."

It is a triumph of technology. Man-made moons circle the earth and serve as bridges for information, helping to connect countries and continents. Perhaps at this very moment they are relaying a radio news report, or even a television picture. Echo 1 and Echo 2 are symbols. Admired by millions as they make their shining orbit across the heavens, they provide the connecting link between almost incomprehensible space exploration and the man on the street.

The great display failed to take place that night. The few stalwart individuals who ventured outside warmly clothed, waited in vain. At the moment of their meeting, Echo 1 and Echo 2 were in the earth's shadow. No sunlight reached them, and they remained hidden from the eyes of the observers.

The prehistory of these "balloon" satellites goes back to the year 1944. In the last months of the war, the Germans directed a radar apparatus of the "Würzburg" type at the moon and sent short radio impulses toward it. A few seconds later they were able to register the reflected signals on their receivers – weak, but nonetheless perceptible. For the first time a body outside the earth had been successfully utilized in the transmission of radio signals. No one worried about the quality of the transmission. After all, they were standing at the beginning of a revolution. Marconi's radio transmissions in December, 1901 from Poldhu in Cornwall to Newfoundland only consisted of a weak "peeping". Think of the development since then! The entire communications technology had been revolutionized.

Because of the confusion of the post-war years, radar experiments could not be continued in Germany. However, they were resumed by the Americans and the British. The American Army Signal Corps in particular had been trying with great success from 1946 on to utilize the moon as a reflector for attempts at transmission. Although the moon reflected no more than seven percent of the transmitted energy that reached it, not only were radio communications established in this way, but even photoradiograms were transmitted. In 1958 the first radio contact between Fort Monmouth in the United States and Bonn, Germany took place.

Today we smile at these attempts. It seems absurd to use the moon for radio transmissions, since it is 250,000 miles from the earth and has basically unsuitable reflecting characteristics. What was it like, however, in the fifties? In the interim this has almost been forgotten, but an entirely different situation prevailed then. More than a million telephone calls were made across the Atlantic in 1950, but under what conditions! There were as yet no telephone cables. All transcontinental phone calls had to be sent via short-wave radio. It was known that radio transmissions could be made with far

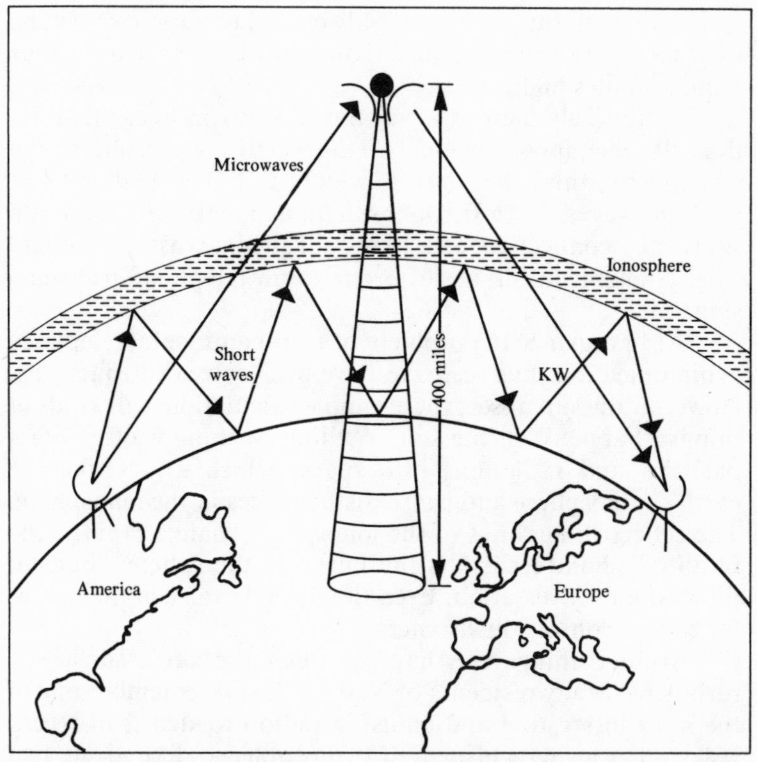

Microwave transmission across the Atlantic is only possible with the use of relay stations. To get by with only one tower would mean that it would have to be 400 miles high. Shortwave radio takes advantage of the fact that certain layers of the ionosphere reflect short waves.

less distortion in the microwave range, but radio waves only spread out rectilinearly, like visible light. Sender and receiver have to be in "line of sight". Microwave transmission over long distances is only possible if the signals can be "passed on" by several relay stations. But where can such stations be built in the middle of the Atlantic? The transmissions systems on the continent require, for example, intermediate stations every twenty to thirty miles. If only one relay station were to be used for a transatlantic transmission, it would have to be four hundred miles high!

Fortunately there is a way out of this complicated situation. In the upper regions of the earth's atmosphere, the ionosphere, there are strata which reflect the short and medium waves. With the ionosphere as a natural reflector, a shortwave connection can be established across the ocean. This phenomenon is basic to our radio telephone transmission.

At first glance the problem of transcontinental telephone communication thus seems to have been solved satisfactorily. However, the sun upsets these simple calculations; all kinds of things "happen" on the sun. Without warning it often hurls particles and radiations into space, which also strike the earth's atmosphere and cause disturbances in the ionosphere. The reflective qualities of the ionosphere change rapidly; the radio telephone transmits the music of the spheres, but not the desired conversation. Even the friendly voice of the operator cannot console the listeners.

Strange things can happen when the sun's surface is turbulent. Many residents of New York still remember one of the most interesting and amusing radio broadcasts of recent years – the loving whispers of two people, so deceptively real that no one would have thought it a radio broadcast. Indeed, it was not, and the two people concerned were less than enthusiastic when they learned that, because of a disturbance in the ionosphere, their intimate telephone conversation had been picked up by ordinary radios.

Because of these inconveniences other means were sought to facilitate transoceanic telephone communication. Thought turned toward the transatlantic telegraph cable, which had been successfully laid in 1858. This original cable soon failed, but was replaced, and telegraph communication boomed in the following years.

The main difficulty was in developing suitable underwater amplifiers; without them the spoken word was so severely deadened that developing a usable telephone cable seemed utopian. The problem was solved during the fifties. In 1950 an experimental cable was laid the short distance from Key West, Florida to Havana, and in 1957, there followed the laying of the first transatlantic telephone cable, TAT 1, at a cost of $25,000,000. It was capable of carrying thirty six clear telephone connections simultaneously.

TAT 1 was of course only a drop in the bucket; telephone communications developed at a frantic pace. Through improved communications the continents were brought closer together. Commerce had expanded to an international level and, consequently, the business world demanded fast, reliable telephone connections. By 1960 the number of transatlantic calls had risen to almost four million; and this was achieved with just one cable of minimal transmission capacity. People still relied primarily on radio transmission, and the disturbances and disconnections in these calls continued.

The situation finally changed in the sixties, not only because of the use of satellites, but also, even if only in small part, because of new cables. After TAT 1 came four more transatlantic cables, TAT 2 to TAT 5. TAT 5, laid in 1970, could handle 720 simultaneous telephone calls. Without the communications satellites, however, the flood of transoceanic telephone calls could never have been mastered. The number of calls across the north Atlantic increases yearly by ten to fifteen per cent, and across the Pacific by even more.

At the end of the fifties the business world might have resigned itself to the fact that communications would not always be as free of disturbance as might be wished. The military, however, was not satisfied. They had to have absolute reliability, and could not wait for the laying of more cables, for example, in the Pacific. Therefore they worked on using the moon to communicate between very distant points. It made little difference that contact could only be made a few hours per day – when the moon was in sight of both the sending and receiving stations. At least they could work in the microwave range, and did not need to worry about disturbances in the ionosphere. In 1960 a regular teletype and visual transmission communications system between the American Navy headquarters in Washington and Hawaii was established via the moon.

The moon was of marginal interest for civilian communications. What was needed were telephone circuits, and when these were made via the moon, there was always a five second wait for an answer. That was frustrating. This five second delay comes about because the speed of all electromagnetic waves, including radio waves and visible light, is finite, even though these waves have the extremely high velocity of 186,000 miles per second. At this speed the transmission time of a signal from the earth to the moon and back is about two and one-half seconds. In a telephone conversation both the question and the answer are conveyed via the moon. That adds up to five seconds. For this reason, using the moon as a means of communication has remained purely in the military domain.

In the winter of 1959–60 the residents of Sugar Grove became aware of just how military it was. At that time, Sugar Grove was a small spot on the map, with thirty two inhabitants, situated in a remote section of the Allegheny Mountains in West Virginia. It was rare that an outsider wandered into it. Time seemed to have passed by this part of the United States, but that changed quite suddenly.

That winter the first work crews appeared unexpectedly in a neighboring valley. Peace and quiet were gone. Machines rattled continuously, and powerful excavators dug deeper and deeper into the virgin soil. Orders echoed back and forth, chains clanked, columns of trucks carried the rubble away. The inhabitants of Sugar Grove only had one topic of conversation: the new building site. What did it mean?

Soon rumors began to spread. The Navy was building a powerful radio telescope, 585 feet in diameter: much larger even than the recently installed "giant ear" at Effelsburg in the Eifel Mountains. (With its diameter of 325 feet, this telescope, more than ten years later, is the largest of its kind.) It was also strange, since two radio telescopes were already operating near Sugar Grove. What was the Navy really doing?

The puzzle was never solved unambiguously. There are, however, clues to a probable explanation. Closer scrutiny revealed that the American intelligence agencies, the CIA (Central Intelligence Agency) and NSA (National Security Agency), were behind the project. Open discussion of the radio telescope was allowed only behind closed doors, and only at the very top, in the National Security Council of the United States. Among other things, the NSA monitors communications in the eastern bloc.

Indeed, the solution to the puzzle probably lay here. Around 1960 there was still no way to monitor Soviet military radio communications in the UHF range. Due to the rectilinear propagation of these waves, these communications could not be picked up beyond Russia's borders. The UHF waves did, however, reach the moon undisturbed, and were reflected back to earth. With the aid of a gigantic radio telescope the signals from the Soviet Union could be received in the middle of the USA, at least when the moon was in the proper location.

The Sugar Grove telescope was never installed, however. Robert McNamara, then Secretary of Defense, personally tabled the project in July, 1962. Construction was halted. One of the most expensive dumps in the world had been created. A few days earlier, on June 18, 1962, the first American electronic reconnaissance satellite, "Ferret", had been launched. It was capable of performing the same functions as the radio telescope, but more efficiently. Was it coincidence that both events followed each other so closely?

Before the Sugar Grove project was begun, and even before the first man-made earth satellite, Sputnik I, was shot into orbit, the idea of creating an "improved moon" arose. This moon was to be the cornerstone of an international communications system which would satisfy man's growing hunger for information and the exchange of ideas. It is true that the first telephone cable had already been laid from Key West to Havana, but it was foreseeable that cable technology would not be able to meet all the demands placed on it in the years to come.

Echo 1

There were two ideas behind the plan to improve the moon. First of all, the moon is a poor reflector. A man-made moon in the form of an almost perfect sphere could be equipped with a metallic surface which would reflect back to earth almost 100% of the radio emissions which hit it. As previously stated, the moon reflects only 7%. To be sure, only a fraction of the radiated energy would strike the man-made moon, thus decreasing its effectiveness, but that is a general problem with all reflectors, including the natural moon. Careful focusing of the rays can hold this within tolerable limits.

The second idea was to bring the moon closer to the earth, so that the transmission time of signals from earth to moon and back would be short enough to allow a reasonable conversation. On the other hand, the man-made moon should not be too close to the earth, as the "line-of-sight conditions" would otherwise be too unfavorable. For use in transatlantic communications it must be a minimum of 400 miles high. This was exactly the height necessary for a relay station in the middle of the Atlantic, if it were to receive microwave signals directly from North America, as well as from Europe.

The idea for such a man-made moon was first advanced in 1955 by John R. Pierce of the California Institute of Technology. In 1958 Pierce approached those responsible in Washington, where he emphasized the sharp rise in communications. According to the most recent figures available to him, the number of intercontinental telephone calls had risen from eleven thousand in 1927 to three million in 1957. This flood could be mastered only if satellites were used to receive signals from earth and to reflect them like "echoes" to other stations. Thus, not only a new name was coined, but also the cornerstone was laid for a development which meanwhile has come to be viewed as inevitable. With a few exceptions, however, the balloon satellites, especially Echo 1 and Echo 2, have become museum pieces.

Nevertheless, they have a very special significance. Due to their function as reflectors in the sky, they had to be larger than any previous satellites, so that on clear nights they traversed the heavens glowing brightly. Newspapers printed daily notices and directions announcing when they could be observed. In this way, they became the messengers of a new age.

I still remember very clearly a warm summer night in 1965. I was standing above Lake Lugano at the foot of Mount San Salvatore and was enjoying my surroundings – the view of the lake and the lights of Lugano and all the other small towns on the lake. Unexpectedly, balloon satellite Echo 2 appeared in the night sky and moved across the glittering sea of stars, glowing majestically. It was an unforgettable experience.

In 1960 the construction of a man-made moon suitable for communications proved to be extremely complicated. It was quite clear that it had to be a balloon, because the strength of a reflected signal increases with the square of the surface of the reflector. Rockets were not yet capable of carrying heavy, bulky satellites. A balloon offered the advantage of being small and light when collapsed, so that it solved the transportation problems. When blown up, it fulfilled the conditions set by the communications technicians.

It was difficult to select the material for the envelope. It had to be resistant to tearing and as sturdy as possible. It also could not be too heavy, despite a planned diameter of about 100 feet. In addition, good reflecting characteristics, such as only metals exhibit, were essential. It seemed almost impossible to satisfy all the requirements. Nonetheless, they had to be insisted upon. For example, if the tensile strength were reduced to lessen the weight, it would of necessity mean that striking meteorites would make large holes in the balloon, causing it to collapse.

Finally, they decided to make the envelope out of a very sturdy synthetic foil – Mylar. As improbable as it may sound, this foil was thinner than the cellophane wrapper on a pack of cigarettes. Its cross section was no more than 0.0032 inches! An aluminum layer of only 0.0006 inches, which was sprayed onto the outside, provided the reflecting properties. The aluminum reflected up to 98% of the microwave signals, as well as those in the range of visible light. The entire envelope was so thin that, when held up to a light, it seemed almost transparent, and the 100 foot wide balloon weighed little more than 135 pounds.

The technicians had had a most unusual idea for inflating the balloon while in orbit. They put a mixture of ten pounds of benzoic acid and twenty pounds of anthraquinone in the collapsed balloon. As soon as the balloon was ejected from its 215 inch protective capsule, the benzoic acid vaporized, because of the warming rays of the sun, and the satellite attained its taut, round shape. The anthraquinone, on the other hand, vaporized only slowly, so that the gases which gradually escaped through small meteorite holes were continually replaced. Before the launching, the balloon was inflated and examined from stem to stern. Using captive balloons, the technicians rose to the sides of the future communications satellite. There was no other way to examine its entire surface. When all the tests were finished, the balloon was deflated and folded up. Finally the great moment came. On August 12, 1960 Echo 1 achieved its orbit around the earth. A new chapter in the history of communications technology began.

What a beginning! The record of the experiments performed on Echo 1 shows how enthusiastically the communications technicians attacked their job. A new field of operations lay before them. A man-made moon circled the earth. They now had to prove that it would be indispensable in the future.

During the satellite's very first revolution around the earth the scientists at Goldstone, California aimed their large antenna at this new reflector in the sky. At a power of ten kw, they broadcast a pre-recorded message of 127 words from President Eisenhower. Ironically, this was the same president who had worked so hard against the space program during his entire term in office. The reception in the Bell Laboratory in Holmdel, New Jersey, on the East coast, was clear and distinct. With this test the superiority of transmission via satellites over radio transmission was conclusively demonstrated: at the very time that Eisenhower's voice was being received, normal radio connections had been interrupted by a violent solar eruption.

August 18, 1960. The first transatlantic connection is established. This time Holmdel is the transmitting station. A telegram sent from here is received at practically the same moment by the French station, Issy-les-Moulineaux.

August 19, 1960. Echo 1 can relay not only telegrams and voice transmissions, but also pictures. Associated Press takes part in the experiment. AP has made its phototelegraphic equipment available. In the station in Cedar Rapids a picture is scanned line by line, sent into space, reflected by Echo 1, and picked up by a receiving station in Richardson, Texas. The transmission takes place with the same speed as in normal phototelegraphy. After less than five minutes the receiving station has the finished picture.

August 22, 1960. The first transatlantic conversation is held via Echo 1. Again, the message is received clearly and distinctly in Europe. It is a description of the satellite experiment. The text was chosen with great care. Nothing is left to chance. The time is past when, in similar circumstances, children's songs became famous, for example "Mary had a little lamb" – the first words ever taped.

In the first month after launching, about four hundred experiments were carried out with Echo 1. Messages were sent by teletype and telegraph within the United States, but also to England and France; telephone conversations were held. Even phototelegraphy was not ignored. The Postal Service performed its own experiments. On November 10, 1960 a public announcement reached Newark, New Jersey. It asked people to mail their Christmas presents early. The message was sent from Stump Neck, Maryland, and broadcast via satellite.

Even in the Federal Republic of Germany experiments were carried out in connection with Echo 1. The public observatory in Recklinghausen transmitted the words "Test Echo 1". Lacking a precisely tracked radio telescope, they mounted a radio antenna on an optical telescope and focused it on the satellite. It's all a question of having imagination! The test words were received in Southern Westphalia.

However, with time, reception grew poorer. Echo 1 became prematurely "senile". Apparently it was hit by meteorites more often than was anticipated. They destroyed two square inches of its expensive surface daily so that the gas inside escaped, and the envelope lost its tautness. The satellite collapsed into shapelessness. To top it all off, there was a collision – probably the first satellite collision ever. Echo 1 collided with its launching canister, which had been following an almost identical orbit around the earth.

In spite of this, the satellite once more made worldwide headlines on April 24, 1962 when the first successful television transmission was made via satellite. This was just in time; only a few weeks later the launching of Telstar 1 made it clear that, in the long run, balloon satellites, although a milestone in communications technology, could not compete with their active "brothers".

Television pictures were transmitted from Camp Parks, California, reflected by Echo 1, and received in Millstone Hill, Massachusetts. The most famous of these shows nothing more than the letters "M.I.T.", the abbreviation for the Massachusetts Institute of Technology. It was extremely distorted and only moderately clear due to the changes in shape that the satellite had undergone.

Finally, after seven and one-half years, and over forty thousand orbits around the earth, Echo 1 entered the denser regions of the earth's atmosphere on May 24, 1968 and burned up south of Hawaii. It was mourned by the romantics, for whom a star passed away, and left a trace of nostalgia among the technicians who were already moving into a new world of communications technology.

Although with the appearance of the first active communications satellites, the age of reflectors in the sky was essentially past, the "Echoes" did not disappear completely from the scene. They no longer played an important role, but now and then they reappeared suddenly and made news. Because of their specific characteristics, they were valuable for surveying the earth, for investigating the upper atmosphere, and for determining the force of the sun's radiation. As astounding as it may seem, this extremely minute force has a noticeable effect on the orbit of a satellite – the larger and lighter the satellite, the greater the effect. The force of this radiation can only be determined with great precision through the use of balloon satellites.

After the launching of Echo 2 only the military devoted themselves to the problem of communication reflectors. As though they had not recognized the signs of the times, they produced a satellite in 1966 which can only be designated a "living fossil". It certainly exhibited very amusing properties. The designers had worked hard to produce something unique. The primitive sphere had given way to a wire frame covered by a new kind of envelope. Once in outer space, the wire frame unfolded to form a spherical structure. After a short time the envelope disintegrated as a result of exposure to the ultraviolet rays of the sun. What remained was the wire frame, which, because of its construction, could reflect incoming signals more accurately and efficiently than the Echo satellites. In addition, meteorites could not destroy a satellite of this type so easily, and it offered less resistance to the thin atmosphere and the radiation from the sun, so that its orbit was more stable. Nonetheless, it was a step backwards, coming one year after the launching of the electronic wonder "Early Bird" – the satellite which gave the western world a brilliant display of international television.

An experiment which the American military carried out in the early sixties is more understandable; at that time there were no active communication satellites. Moreover, it is a classic example of the fact that, even in a democracy, military priorities always take precedence over civilian priorities. With reference to the United States it can be more accurately stated: first comes the military, in second and third place comes the military, and then after a large gap comes everything else. That is because in the United States, the Army, the Air Force and the Navy are not only kept strictly separate, they often work against each other, so that they must be considered separately.

It was thought that the more satellites there were in a communications system, the more efficiently it would function. A network of satellites is of greater value than a single satellite, which can only provide a connection between two places for a short time. Why shouldn't one position a ring of reflectors around the earth? Not much time elapsed between the idea and its implementation. In October, 1961 the Americans launched an early warning satellite, Midas 4, which contained a load of copper needles in addition to its primary equipment. Once in orbit it ejected its load, and 350 million tiny copper needles were released. A cosmic experiment was thereby started whose effects were not localized, and whose consequences were completely unpredictable.

It was the astronomers who had first raised the strongest objections. No one knew to what extent a ring of copper needles around the earth would disturb their observations; no one knew what negative effects the experiment could produce, especially for radio astronomy. Nobody consulted the astronomers, however. The objections of the scientists faded away unheard. The military was stronger. They were not strong enough, however, to conjure up success. The copper needles did not even begin to spread out. The astronomers relaxed.

In the meantime a second, similar experiment was prepared. In May, 1963 it was ready. Again, a packet of copper needles was flown into outer space, again it opened, and this time the experiment was a success. After about fifty days the earth had a man-made ring of tiny dipoles – nine miles wide and nineteen miles thick. The military had gotten its way.

Fortunately, the negative effects proved to be less significant than the scientists had feared. The work of the astronomers remained almost unaffected. It could have been otherwise, and probably no one would have been called to account. The stronger party always gets its own way.

It became clear in 1958 that experiments of this magnitude do not always have such mild consequences. It had just been proven that the earth was surrounded by radiation belts which held electrically charged particles captive, like a trap. These belts were named the Van Allen belt, after their discoverer. These radiation belts provided the basis for an experiment which was to have consequences for years to come.

One year earlier, the physicist Christofilus had carried out theoretical experiments which exactly predicted the following effects: electrically charged particles of low energy which strayed into a certain area of the earth's magnetic field, would be captured and held. Undoubtedly, he had no idea that the verification of his calculations lay close at hand. Perhaps he did not recognize the consequences of his theories, either. However, it was immediately clear to the military what sort of tool had accidentally fallen into their hands. It was only necessary to create an artificial shower of electrically charged particles at a great altitude, in order to possess the best weapon against approaching missiles. An exploding atomic bomb was considered a good source of such particles.

In August and September, 1958, when three American atomic bombs were exploded at short intervals about 310 miles above the earth's surface, the experiments were still shrouded in secrecy. No one connected these explosions with increased reports of various natural phenomena. In northern America, near Norton Sound, where the bombs had been launched, disinterested observers enjoyed a colorful spectacle – an irridescent polar light lit up the sky. A second polar light, this time in the form of the aurora borealis, lay over the Atlantic. An unusual atmospheric illumination could also be seen in other parts of the world.

It meant absolutely nothing to the experimenters that they had succeeded for the first time in artificially creating an aurora borealis. They saw only that the experiment was successful. A man-made radiation belt had been created. The success was dearly bought, however; the natural condition of the upper atmosphere was severely disturbed for years to come by the experiment. This was especially deplorable at a time when a new science had just been born, a science which enabled man to investigate more closely that part of the world which had previously been unknown and unexplored. We must agree with the Nobel Prize winner Max Born, who once said of space exploration, in another context: "A triumph of the intellect, but a tragic mistake of reason".

Fortunately, such occurrences are rare. It is very regrettable, however, that even with minor experiments no consideration is given to the effects on the environment. The natural condition of the atmosphere changes daily. So far it is not clear how these changes affect the balance of nature. Let us consider only the upper atmosphere: every launching of a space rocket, every launching of an intercontinental missile changes the electrical properties of the atmosphere, even if to a far smaller degree and on a far smaller scale than did project "Argus" which was just described.

Both the Soviets and the Americans focused attention on

this fact as early as the fifties. However, there were also those who welcomed Project Argus, because it provided a fantastic possibility for "observing" the launching of foreign intercontinental missiles, and of expanding the classical early warning systems which were felt to be so essential during the cold war. In principle, this ingenious system amounted to the following: radar sends out radio waves which graze the ionosphere and are picked up by a radar receiver thousands of miles away. Launching a rocket disturbs the ionosphere, so that its characteristics change. These changes will then be apparent in the received radar signal.

The realization of this idea proved to be extremely complicated, since the problem of distinguishing natural fluctuations in the ionosphere from artificially produced ones had to be solved first. A series of experiments was run until, in 1964, the development of the first OTH radar system was begun. OTH is an abbreviation for "over-the-horizon", and means that the area of observation lies beyond the horizon. Finally, in 1968, NORAD, the North American Air Defense Command, began to use this system on an experimental basis. Recently, it has been in full use. The system has the sole function of monitoring the launching of intercontinental missiles in the Soviet Union. Its transmitting stations are located in Southeast Asia, while its receivers are stationed in Europe – one of them in the Federal Republic of Germany!

Inspecting the balloon satellite, Echo 2. Structural imperfections are visible only when the balloon is inflated. Here, a technician uses a captive balloon to inspect the entire surface.

2. Private Industry Enters the Picture

The Transition to Active Satellites

November 25, 1963. The hands on the clock seemed to be moving very slowly, as though they were trying to increase the tension. The picture on the television screen flickered. The moment had finally come. The picture stabilized, and the words "via Relay" appeared. It was 11:36 EST. In Japan the new day had barely begun. Only a few people were sitting in front of their television sets to watch what was happening: the funeral ceremonies for President John F. Kennedy, who died so tragically. It was an historic moment. For half an hour Japan and the United States were directly connected via television. The Japanese were able to feel the sadness surrounding the death of the U.S. president, as if there were no ocean separating the two countries.

A few days earlier NASA had completed its first successful test broadcast via the satellite, Relay 1. On November 22, the day of Kennedy's assassination, Japan received the first television program sent from America. Because of the importance of the funeral, the American technologists made their test earlier than planned. A chain reaction began. It was a unique situation; from this time on, cause and effect were reversed. International political events no longer motivated such complicated communications technology, but rather, mastery of the technology led to regular transmissions.

Kennedy's funeral was broadcast to the four corners of the world. It was seen live in Europe and North Africa, as well as in Japan. The twenty three countries of Eurovision picked up the program. Even the eastern bloc did not want to be left out. Seven countries of Intervision, the east European equivalent of Eurovision, also joined the broadcast. For a short moment the death of a president united the world.

What led to this impressive demonstration of modern television technology? Long before the first balloon satellites were used to relay communications, farsighted scientists had realized that the future of communications technology lay in active repeater satellites. A balloon could reflect any type of signal, but because of the long relay distance, and the scattering effect of the sphere's surface, the original signal was highly distorted. It reached the receiving station weakened many hundred million times, and could only be made intelligible at great expense. The demands placed on the relay medium increase in direct proportion to the amount that has to be relayed within a given period of time. It makes a difference whether what is broadcast is only a few Morse code signals or a clearly understandable conversation, with all the nuances of human speech. The transmission of pictures is even more complex. Broadcasting a television program is only possible on a broad band. In the mid-sixties the capacity of all transatlantic cables taken together was not sufficient to relay a single television program.

Although a television transmission via the balloon satellite Echo 1 had succeeded on April 24, 1962, the ground stations had had to exert considerable effort, and the reception had only been mediocre. Apparently the quality of the picture could be improved only slightly. However, active communications satellites offered a solution. Active satellites do not reflect a signal; rather they receive it, amplify it many times, and then rebroadcast it on a different frequency. The solution to the problem lay here: the active amplification of the signals by the satellite itself. Reception was much clearer, and external factors, such as Echo 1's distortion, which adversely affected transmission quality, were eliminated.

The Army Signal Corps quickly recognized the signs of the times. On December 18, 1958, it launched the first test satellite, Score 1, and led the way to a communications network that spanned the globe. As with many satellite names, Score is an abbreviation. It stands for Signal Communication Orbit Repeater Experiment. The name characterized the satellite's activity – it received signals from the earth, stored them on tape, and rebroadcast them to the earth on command. The ground stations in contact with it were in Texas, Arizona, and Georgia.

Interestingly enough, Score was immediately attacked as a provocation by the eastern bloc. First, it was a military project, and second, it was able to spread news from the West – even if only weakly. This is something which the East does not like at all. Their aversion can be traced throughout the entire history of space exploration. Even today free communication via satellite is a thorn in the side of the Soviets. Indeed, the Soviets consider this a greater problem than unlimited observation and surveillance by the satellites of foreign countries. It poses a threat to their ideology! What will happen if the Americans orbit a communications satellite blanketing Soviet households with radio and television broadcasts of a capitalist nature? Will the Communist system be able to withstand this influence?

However, the technology of satellite communications had not reached this point in 1958. Only large, complicated antennae could receive the broadcasts, and to speak of provocation was ridiculous. Score's main task consisted of broadcasting a taped Christmas message from President Eisenhower for thirteen days, day after day. Finally, after thirteen days, the satellite's batteries went dead, and it became silent. A few days later it burned up in the upper regions of the atmosphere.

Score was a primitive beginning. But that is the way things are – even a house is not built in one day, but rises brick by brick. What starts as a crude foundation is one day a masterpiece of architecture. We can trace an unbroken course from Score to those first transatlantic television broadcasts that held 200 million viewers spellbound. This sequence is not even very long. The final link is Courier 1B, another satellite of the Army Signal Corps, another "courier in space": it could receive, store, and rebroadcast messages at any time.

Courier 1B had one structural characteristic which all later communications satellites also possessed. Its source of energy was no longer chemical batteries, but sunlight. Solar cells fixed to the exterior converted the sunlight to electricity. The life of the system was greatly increased, since it was no longer limited by batteries.

Courier 1B was inaugurated with great fanfare. The first broadcasts soon after launching on October 4, 1960 revealed a bit of American sentimentality, as the satellite broadcast the Constitution of the United States to the earth. Everyone could see how proud the Americans were that they, and no other nation, were the pioneers in communications satellite technology. A similar action later led to heated arguments when the astronauts Armstrong and Aldrin planted the Stars and Stripes on the moon. Apparently that too is a part of space exploration. After all, are there not Soviet pennants and emblems on the moon, Venus, and Mars? Didn't Luna 10, the first satellite to orbit the moon, in April 1966, broadcast the "Internationale"? What did the first Chinese satellite broadcast to the whole world in April 1970 but "The East is Red"?

Courier 1B showed its actual capabilities in a different way. Because of its orbit, it was only in line of sight of a ground station for a few minutes each revolution. In this short time it had to broadcast the messages it had previously stored on tape. We should also mention that Courier 1B could relay signals directly, when it was within range of both the sending and the receiving station.

Naturally the number of tapes that it carried was limited. Four of them were for teletype and facsimile transmission, while the fifth stored voice broadcasts. Normally not very much information could be transmitted with such limited means. (In comparison, for example, almost ten thousand telephone conversations can be carried simultaneously by modern satellites.) Nonetheless, by means of a gimmick, Courier 1B did become a relay giant. Using a special technique the Americans increased the capabilities of the satellite to such an extent that it could handle 75,000 words per minute, the length of a medium-sized book! A teletype machine needed about 90 hours to transcribe the text the satellite transmitted. In other words, 720 teletype machines working at full speed could make it function at capacity.

Even more impressive than these figures, is an experiment which the Americans carried out to demonstrate Courier 1B's capabilities. At a transmitting station they recorded the 773,693 words of the Bible on a magnetic tape. The process took thirty six hours. How many drinks the speakers later needed for their dry throats went unrecorded. When the satellite came into the range of the station, the tape was played at an accelerated speed. Within fourteen minutes, the 773,693 words reached the satellite, where they were stored. Then Courier 1B disappeared beyond the horizon.

A short time later it appeared above the receiving station. Now the whole process was reversed. Within fourteen minutes the earth again had the text of the Bible. After thirty six hours the technologists were finally convinced: the text had been relayed clearly and distinctly.

Courier 1B's feat was never equalled. Its equipment stopped functioning after slightly less than a month, because of a defect (it had already relayed 118 million words and 60 facsimile pictures). The technology of accelerated transmission disappeared with the satellite. In the long run only a communications system that worked in "real time" was viable. The experience that the technologists had gained from working with the satellite remained.

An Offer from NASA

These experiences benefitted civilian space exploration. Although NASA had concerned itself primarily with the construction of passive communications satellites, it announced in the same year that Courier 1B orbited the earth, that it also intended to start developing active satellites. Simultaneously, NASA made American industry an offer which was unique in the history of space exploration. It promised the greatest possible financial assistance to any private companies wanting to experiment with satellite communications. In other words, NASA was prepared to launch private satellites and to make its tracking equipment available. Thus it entered into a union, the fruits of which all mankind was to enjoy. Naturally it did not do this for free. The enormous expenditure was to be repaid. In addition, every company that wanted to take advantage of this offer had to submit the results of its satellite experiments to NASA. As was NASA's custom, these results would then be available to anyone who was interested in them.

This offer did not have to be made twice. However, not all companies were equally enthusiastic about it; some of them had invested heavily in transatlantic cables, and viewed the communications satellites as strong competition. One thing was clear to them, however – in the long run, they would have to accept them. It was obvious that satellites would show a profit in the near future, and it was better to have a part in this profit than to leave it to others.

American Telephone and Telegraph (AT&T) decided within a few days to accept this magnanimous offer. Its managers could easily estimate that within the next five years the company would need about 5,000 more circuits to handle transoceanic calls; existing and projected underwater cables could not supply this. A satellite system ought to fill this gap at a reasonable cost. Only nine days after NASA's offer AT&T applied to the Federal Communications Commission (FCC) for permission to build its own communications satellite. No satellite could be built or orbited without this permission. The FCC coordinates and supervises the communications industry, which is privately owned. On July 27, 1961 AT&T signed an agreement with NASA which provided for two to four communications satellites in 1962. Meanwhile, experimentation at AT&T shifted into high gear. They had to solve completely new problems. They needed a system of communications satellites which not only accomplished its tasks satisfactorily, but which was also economical. This was at a time when space flight was still in its infancy. The ideal solution had been suggested sixteen years earlier. In October 1945 the English engineer and author Arthur C. Clarke, then Chairman of the British Interplanetary Society, published an article about extraterrestial relays in the magazine *Wireless World.* In this article he investigated the problem of transoceanic communications. He was the first to suggest using active satellites. In an exact analysis he determined that a satellite which was in an equatorial orbit at a height of 23,375 miles would rotate in phase with the earth, and thus would always be above the same spot on the earth. It would be "fixed". Three such satellites orbiting in phase with the earth, stationed over the equator, and spaced 120° apart, could keep almost the entire earth in sight. Only the extreme polar regions would not be in their range. However, since those areas are sparsely inhabited, such a system of three synchronous satellites would be almost ideal for a worldwide communications network.

This system has a further advantage. Since the satellites are fixed, the expense for ground stations in contact with them is quite small. The antennae do not have to track the satellite continually. On the other hand, if a satellite is in low orbit around the earth, it is only in the range of a ground station for a short period. During this time the antennae have to track it with great precision, and even before it has disappeared below the horizon, the next satellite must be visible, in order to maintain uninterrupted communications. A second antenna is necessary to pick up this satellite in time. All of this makes it very expensive.

Nonetheless, AT&T could not take the step toward developing a synchronous satellite system. Technology was not yet far enough advanced. Rocket technology alone caused numerous problems in placing a satellite in synchronous orbit, and the alignment of such a satellite, as well as its stabilization, proved very expensive. Furthermore, even a synchronized satellite does not remain firmly fixed in position. The smallest disturbances, for example, the unequal distribution of the earth's mass, will cause it to move out of its orbit. Only precise corrections in course at definite times can compensate for this motion.

The extreme opposite of a system of synchronous satellites is one in which the satellites orbit the earth in arbitrary, low orbits. At any given time there is bound to be at least one satellite in the right position for communication between ground stations. If there are enough satellites, a second satellite will appear over the horizon before the first one disappears. With this system it is not so important that a satellite attain a specific orbit. Such a system makes fewer demands on technology. This was the system which AT&T chose.

It was estimated that 50 to 60 satellites in low orbits, or 24 satellites at an altitude of about 6,250 miles, were necessary to maintain uninterrupted international communications. Nonetheless, in comparison to a worldwide cable system of equal capacity, there would still be an advantage! This was true, even when they added in the expensive ground stations that had to be built. A subsequent study showed that, to open 20,000 communications circuits, a synchronous satellite system with 100 ground stations would cost $277 million. The cost would be ten times greater, $2.75 billion, for a satellite system orbiting at an intermediate altitude, while a system of transoceanic cables would cost about $4 billion.

When the men at AT&T responsible for the project arose before sunrise on the morning of July 10, 1962, they were haunted by these figures. Among these men were John R. Pierce, inexhaustible in his pursuit of new ideas and already famous for his connection with the Echo satellites, and Frederick Kappel, the Chairman of the Board of AT&T, who

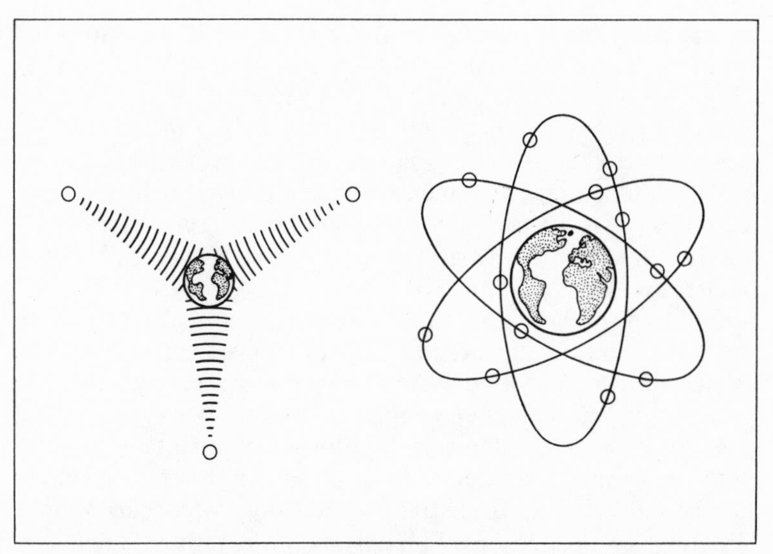

The above diagram shows two alternatives for a worldwise communications network using satellites. Left: three satellites in synchronous orbit are sufficient to reach almost the entire inhabited world with communications. Right: the same results can be obtained by 50 to 60 satellites in low, static orbits.

awaited the approaching debut in the ground station at Andover, Maine. Would everything go off all right? Would the first private satellite in the world, Telstar 1, fly around the earth safely in orbit?

Telstar

The countdown at Cape Canaveral (for a long time called Cape Kennedy, but now again known by its original name) went off without any complications. The decisive moment came at 4:35 Eastern Daylight Time – majestically the powerful Thor-Delta booster rocket rose from the launching pad, and accelerated until it was lost to sight. The scientists began to breathe; they rejoiced. It had been done. The first orbital figures showed that Telstar 1 circled the earth at an altitude of from 514 to 3,040 miles. This was high enough to establish contact between America and Europe for one half hour.

First however, the satellite's systems had to be checked out. Failure of one part was not necessarily catastrophic, since the most essential components had been duplicated. If a part actually did stop functioning, it could be replaced immediately. The satellite, which had a diameter of only thirty four inches and weighed only 170 pounds, had about 15,000 parts – including 1,064 transistors, 1,464 diodes, and 3,600 solar cells to generate the necessary power! The tests were completed during the fifth orbit: telemetry was clear and distinct, and the experiment could begin.

Suspense grew. On this day, for the first time in the history of mankind, a television program was to cross nature's barrier, the ocean. Even Echo 1 had not done that. Would Telstar 1 do its job? Would it earn its designation, "television satellite"? Actually this satellite and its successors do not really deserve the designation "television satellite", because relaying television programs is only one of their functions, although perhaps the most spectacular. Sixty two-way telephone conversations could be held simultaneously via Telstar 1, and television only took up 2% of the capacity of the later INTELSAT satellites.

The real importance of communications satellites lies in expanding international commerce. The rates for transatlantic telephone calls have fallen drastically, and that has led to a sharp increase in telephone communication between America and Europe – an increase which has positively affected all branches of international economy. Today, in the age of direct long distance dialing, a conversation across the Atlantic is no longer something unique. The countries of the world are closer.

Maybe this was in the mind of the man who picked up the telephone at the ground station in Andover on July 10, 1962. It was exactly 7:28 P.M. Eastern Daylight Time when he spoke these prosaic words: "This is Fred Kappel at the ground station in Andover, Maine. I am calling via our satellite Telstar. . . how do you hear me?" The answer from Vice President Johnson at the White House was just as prosaic: "You're coming through fine, Mr. Kappel."

Telstar 1 had passed its first test. Now they had to see whether television transmission would work. Just a few minutes later the large Radom at Andover came through. This is the sphere that protects the broadcast antenna from the weather. The U.S. flag was waving in the foreground, while the "Star-Spangled Banner" played in the background. Again, Europe was an eye-witness to the pride that America felt at this moment. Television screens at the ground stations in Pleumeur-Bodou in France and Goonhilly Downs in Great Britain lit up with exactly the same picture, although somewhat blurred, at least in Goonhilly Downs. The giant step across the ocean had been made.

During the next few minutes they saw those who had been instrumental in bringing this project about. Then the satellite itself appeared. After 25 minutes the apparition disappeared just as suddenly as it had appeared. Telstar 1 had sunk below the horizon. An historic transmission was over – a transmission that was just as meaningful as the first telegraph signal that crossed the ocean in 1901 and started a technological revolution.

Several times in the next few days television signals went back and forth over the Atlantic. The second evening America had the pleasure of enjoying a program from Pleumeur-Bodou – eight minutes of France. After a few words by the French Postmaster General, Jacques Marette, they showed an excerpt from a Paris variety program. It had been recorded only a few hours earlier, and then rushed to the broadcast station: a bit of culture. The English got their chance on the third evening, with a look at the control room at Goonhilly Downs, sent via satellite. Three different evenings brought three telecasts from different countries. Even the choice of programs showed the different mentalities involved. Wasn't this an early indication of how much communications satellites, properly used, could contribute to international understanding?

In its first days Telstar 1 had proved so successful that those in charge decided to show the entire world exactly what they could do. The continents were to be linked by a live television broadcast. On the evening of July 23 the television cameras ran: in Seattle and Rome, in New York and Duisburg, in Vienna, Sicily and Cape Canaveral, in Paris and London, in Geneva, northern Sweden, and along the Canadian border. It was a kaleidoscope of pictures and impressions: from the President to an Indian chief, from a space flight hangar to a power plant in the Ruhr. All of these impressions were sent live via Telstar to the U.S.A., Canada, and sixteen European nations. The number of people viewing this highly acclaimed "monster" show was estimated at 200,000,000. Telstar became famous overnight.

Other experiments were being conducted with the satellite. Telephone conversations, facsimile pictures, teletype messages, and all kinds of data crossed the Atlantic. In every instance the new communications bridge in outer space showed its usefulness. Telstar functioned with the greatest precision. It was equal to the most extreme demands, as General Electric demonstrated early in 1963.

General Electric had just developed a device which permitted transmission of certain data from one computer to another. Communications between machines – this was a step into the future. At that time it was something quite new; today it is routine. This step led to computers which are so big that it is difficult to make full use of them, that is, to use them only for those problems which smaller computers cannot handle. Small research groups, however, cannot afford a large computer which they will need only twice a year. Therefore regional research and computer centers have developed over a broad area. These can then tap into the main computer via remote stations. Communication among computers!

In 1963 it was proven that problem-free contact between computers could be established, even across oceans and seas. In an impressive demonstration the output of the computer at the General Electric plant in Phoenix, Arizona was transmitted to Telstar, using the company's new device. Telstar amplified the signal and sent it back to earth where it was picked up in Schenectady, New York and fed directly into another computer. Data had been accurately transmitted a distance of 2,500 miles. This was especially significant, because even one small error in transmission could have resulted in failure. The idea of storing all the world's knowledge in a data bank, so that it is available for everyone is now only a dream. Will it some day be reality? Satellites have contributed their part to solving the technical problems involved; their contribution has already benefitted all mankind.

However, the advances made by Telstar were not welcomed everywhere. Mistrust began to grow, especially in the Soviet Union. The stumbling block was that Telstar was privately owned. The Soviets feared that it would be impossible to exercise any control over its transmissions. Regulating the news, however, is essential in a communist state. The apprehension grew as the first eastern bloc nations, led by Poland, joined the Telstar broadcasts and were sprinkled by "capitalist" television programs.

It might be expected that the Soviet objections had no effect on American television, but that would be far from the truth! Although there was no official American reaction, unofficially it was these protests which caused private television to be muzzled after AT&T launched Telstar 2. Other companies' projects, e.g. General Electric's Mailbuoy, were set aside, or completed by NASA later. At any rate, the semi-private COMSAT was established by an act of Congress at about this time. This was the cornerstone of a purely commercial satellite communications system which was no longer used for experimentation.

Meanwhile, on December 13, 1962, NASA had launched its own active communications satellite, Relay 1. Relay and Telstar belonged to the same generation of satellites. Both had been developed using the data gathered from the Army satellite, Courier 1B; except for external differences, they were similar in construction and capability. The most significant difference was the orbit.

Use of Telstar 1 was severely limited by its low altitude. Even under optimal conditions it was never in the range of both American and European ground stations for more than half an hour. Because of its higher orbit, Relay 1 could maintain a transatlantic connection much longer. It ranged in altitude from 830 to 4,631 miles.

Relay 2, its successor, came no closer to the earth than 1,300 miles, so that with favorable conditions communications could be carried on over the Atlantic for as long as seventy minutes. Of course this was only possible when the satellite was over the northern hemisphere, far from the earth's surface. When its elliptical orbit brought it close to the earth over the northern hemisphere, transmission time sank to only ten or twelve minutes. Since the satellite's perigee gradually progressed along the orbital plane, there were times when Relay (as well as Telstar) was most favorably situated for transatlantic communication, and other times when contact was only possible for a few minutes. They had to accept this variation in operational efficiency.

Relay made the world shrink once again. On January 17, 1963, Brazil became part of the satellite communications network. For the first time television contact was established between the U.S. and a Latin American country. This was a remarkable accomplishment, since Brazil, unlike the U.S. and Europe, did not have the necessary large ground stations.

Fortunately ITT had developed a smaller, moveable station which had an antenna that was only 30 feet across. It could be broken down and transported anywhere in the world. Such a station was taken by truck from Nutley, New Jersey, across the continent to Rio de Janeiro, where it was reassembled. Transmission was good in both directions.

At the time of the first successful television transmission from the United States to Japan on November 22, 1963, there were six stations for receiving communications from satellites outside the U.S.A. – Goonhilly Downs in Great Britain, Pleumeur-Bodou in Franch, Fucino in Italy, Rio de Janeiro in Brazil, Hitachi in Japan and, recently, Raisting in Upper Bavaria. This was the beginning.

There was still no regular exchange of programming via satellite, but rather, because of circumstances, only an occassional transmission. However, in some instances a routine was already developing. This was most evident in medicine. There are several severe illnesses which are so difficult to diagnose that only a few specialists will recognize them. Many realized this, but only the doctors at the Neurological Institute in Bristol acted upon this realization. Using Relay 2, they sent electroencephalograms to the world-famous Mayo Clinic in Minnesota (a unique institution in the field of diagnostics), and obtained the best and most exact diagnoses possible – all to their patients' benefit.

Communications satellites were already being used for medical purposes in a variety of ways. For example, in May, 1965 Early Bird carried a heart operation from the Baylor University Clinic in Houston more than 5,000 miles to a lecture room in a Geneva hospital. In front of his Swiss colleagues, the famous surgeon Michael DeBakey replaced a damaged heart valve with an artificial one. He explained the process in both English and French. Part of the operation was even broadcast over public television, so that several hundred million people all over the world were able to watch.

Although more or less incomprehensible to laymen, the transmission did show quite clearly what advantages the satellites offered. The Swiss medical team had the unique opportunity of observing an authority in their field in person. This was the best way to learn, but it would have been impossible without satellites. Not every doctor can study under the best teacher.

In March 1970 the German Federal Medical Society held its Congress on Continuing Education at Davos, Switzerland. Part of this congress was a forum which took place via satellite, and lasted for many hours. Those taking part in the forum were in Houston, Texas, Davos, and Bad Gastein. In addition, several places in Germany, Austria and Switzerland were tuned in to the transmission. Visual and audio contact were made via satellite. The picture was projected onto huge screens so that everyone, not just the speakers, could participate. More than 10,000 doctors took advantage of this opportunity. Interest was so great that all the tickets were sold out days in advance. The topics discussed were the findings of space medicine as applied to daily medical practice, and advances in the early detection of cancer. The success was overwhelming. Most of the doctors hoped that such broadcasts would be repeated fairly often.

Doctors had also found that communications satellites could be used to bring medical care to isolated areas. Even in heavily populated areas there are not always enough doctors to care for the people. Naturally, remote regions suffer far more from the acute shortage of doctors, particularly specialists. Severe illnesses often go undetected, since there is no way to diagnose them.

In recent years some of these gaps have been filled by well-equipped hospital ships. These are small-scale diagnostic centers and clinics which, being mobile, can be set up wherever there is an acute shortage of doctors. For example, the German hospital ship Helgoland did invaluable work in Vietnam, an area where the ratio of patients to doctors has grown intolerably large because of long years of war.

Hospital ships have truly become irreplaceable, but they

do have their limits. The doctors lack access to extensive libraries and specialists, who are often the only ones able to make a well-founded diagnosis. That will be different in the future. Recently the hospital ship S.S. Hope, sponsored by the privately financed "People-to-People Health Foundation" received special equipment which enables its doctors to talk with the COMSAT laboratories near Clarksburg, Maryland via a satellite of the type INTELSAT IV. Its equipment includes a television transmitter and a camera. The home port of this ship is Maceió, Brazil.

If people learn from this example, and if the system proves a success, then the time is no longer far away, when medical care will be just as good in the remotest areas of the world as it is today in densely populated areas. The problems in disseminating information which have stood in the way of progress will have been solved. By using satellites, specialized clinics will be just as accessible to Vietnam as to Brazil. Experiments with Relay and Early Bird have shown this.

Above: A scene from the live transatlantic television broadcast sent via Telstar 1 on July 23, 1962. This impressive Indian chief could be seen on T.V. screens in the U.S., Canada and in sixteen European countries.

Below: A milestone in the history of medicine. A transatlantic forum of experts takes place with the help of the satellite INTELSAT III. The television pictures are projected onto a large movie screen.

3. The Network is Completed

High in the Sky

In 1972 news of a catastrophe hit the world. An earthquake of enormous proportions had struck Central America. Managua, the capital of the small state of Nicaragua, was almost levelled. Thousands of people were homeless. News of losses and rescue attempts remained on the front pages of the newspapers for days. Finally, one report came through that seemed to be even more important. Like all the other unfortunate people, one of the most prominent residents of Managua, the extremely shy American billionaire Howard Hughes, was also struck by the catastrophe. He had left the country, and had flown to London. For the first time in many years there was a chance to catch a glimpse of this unusual man, the powerful head of Hughes Aircraft Corporation, and was surrounded by more rumors than any of his contemporaries.

It is actually hard to say which deserves more interest: the man himself, or the company whose reins he holds in his hands. In space exploration the name Hughes is almost a magical word. After 1963 virtually all civilian communications satellites were built by this firm, and until a few years ago it was unbeatable in its position. Its success was based on a project started at the same time that AT&T built its Telstar: a communications satellite synchronized with the earth's rotation. Hughes started this project on his own, and later was commissioned by NASA to continue it.

The idea consisted of "fixing" a satellite over the mouth of the Amazon River, so that continuous communications would be possible between the United States and Europe. The satellite was supposed to be as simple as possible. At first, television transmission was not included. Through simplicity of construction the Hughes Aircraft Corporation hoped to keep the weight of the satellite down to 33 pounds. This was a fantastic plan at a time when rocket technology was still in its infancy.

In practice it turned out that the builders had set their hopes too high. The weight of the satellite gradually increased. However, since the power of the rockets also increased, and thus their ability to carry heavier and heavier payloads to a synchronous orbit at 22,375 miles, this did not interfere at all with the project. On July 26, 1963 NASA was able to launch the first synchronous communications satellite in the world – Syncom 2. (Syncom 1 had ceased functioning five hours after its launch.)

Due to small errors in launching, Syncom 2 was really only semi-stationary. A slight inclination to the equatorial plane caused a perturbation in the north-south direction. This was combined with a movement like a pendulum in the east-west direction, caused by the incorrect height of the orbit. The two of these together produced a narrow figure-eight orbit. The center of this eight lay over Africa.

Syncom 2 was used to relay communications of all types, except television, and led a very eventful life.

Many other test satellites in synchronous orbit around the earth shared the same fate. When they were no longer needed at a certain point above the earth, NASA used a radio signal to start their rockets, correct their course, and "push" them where they were needed. The only limitation was that the new position had to be over the equator.

Thus Syncom 2 did not stay over Africa for long. As an American satellite it was too far away from its target to be used in experiments. NASA dragged it back off-handedly and

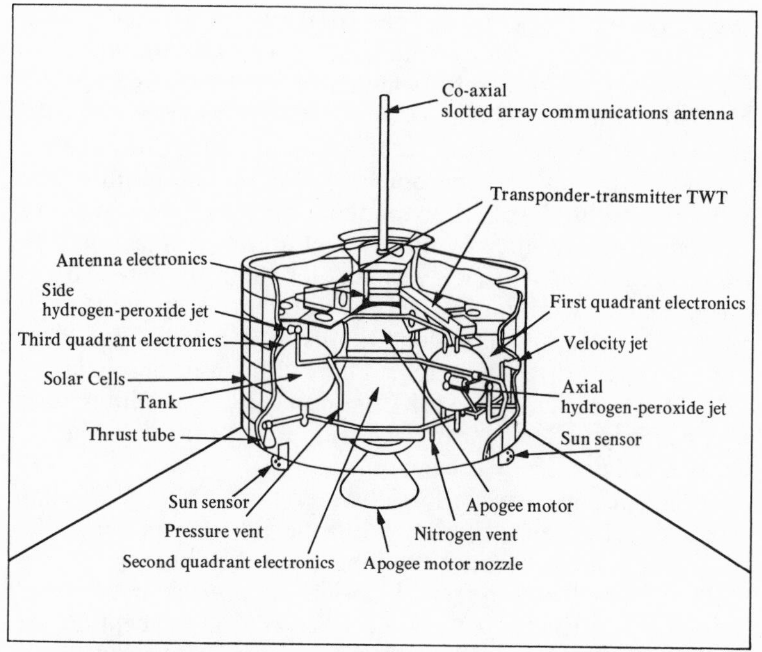

A schematic diagram of Syncom, the "Synchronous Communications Satellite".

fixed it over Brazil. Here, over the mouth of the Amazon, longitude 55° W., it could maintain optimal connections with Europe. When it was no longer needed there, it was pushed to a position over the Indian Ocean.

Syncom 3 led a much quieter life. This satellite was stationed from the very beginning at the intersection of the equator with the international dateline, about 1,250 miles north of the Fiji Islands, in the middle of the Pacific. This was a very central location. It could be reached from the west coast of the United States, as well as from Japan, and this was precisely what the scientists desired. Syncom 3, a technically improved version of its predecessors, was able to relay the

1964 Olympic Games from Tokyo to the United States.

This project almost collapsed at the last minute. The satellite was not successfully launched until August 19, 1964, barely in time for the Americans to watch the greatest sports event in the world. Because of its ability to unite different peoples, Symcom 3 has become a symbol for communications satellite technology. Fortunately only success counts. Although the number of those watching the games on their television sets in America was smaller than predicted, it did comprise many millions. The time difference between the two countries made for many long nights in America. This event was the great breakthrough. The public was enthusiastic. What an experience it was! Communications satellites were necessary, in order to bring the wide world into one's home. This was where the future lay.

The public, especially the American public, started to talk about this new technology. Europe did not join in the discussion as much, because the reports on the Olympic Games still had to be flown in from America. Direct transmissions of the games to Europe (using Relay 2) were the exception. In America, however, the enthusiasm was unbounded. It contributed to establishing a world-wide communications system with the help of satellites. One year later this system was inaugurated with the launching of satellite Early Bird, and today it covers 83 countries. The Soviet Union and China were also showing the first signs that one day they might join the system. Such a decision could have enormously positive results.

COMSAT and INTELSAT

On February 7, 1962 President John F. Kennedy introduced a bill in Congress to create a purely commercial communications satellite system on an international basis. No one could possibly guess to what extent this would affect the existing communications networks. This bill had such momentous consequences that we quote a few words from

Kennedy's original request to Congress: "In my judgment, a new Communications Satellite Act is required to provide an appropriate mechanism for dealing effectively with this subject — a subject which, by nature, is essentially private enterprise in character but of vital importance to both our national and international interests and policies.

"Among the policy objectives pursued in the preparation of this measure have been the assurance of global coverage; cooperation with other countries; expeditious development of an operational system; the provision of service to economically less developed countries as well as industrialized countries; efficient and economic use of the frequency spectrum; non-discriminatory access to the system by authorized users; maximum competition in the acquisition of equipment and services utilized by the system; and the strengthening of competition in the communications industry."

From the beginning there were certain contradictions which were not easy to solve: on the one hand, control of the international communications market is a very important factor in political power. On the other hand, communications in the United States are in the hands of private companies. Where was the responsibility to lie? Probably there would have been many heated debates about this point, if it were not for the fact that there was a great amount of pressure on everybody to finish this international communications satellite system ahead of the Soviet Union. For this reason, on August 31, 1962, the "Communications Satellite Act" was passed. This law called for the creation of a corporation to carry out the necessary organizational work.

Ownership was clearly delimited. One-half of the corporation was to belong to the communications industry, the other half was to be a public company in which any citizen could buy stock. Leadership was entrusted to a board of directors which was elected by representatives of the communications companies, the stock holders, and the government. This is where politics showed its influence.

On February 1, 1963 the new corporation was christened. It was named COMSAT (short for "Communications Satellite Corporation"). Soon after the sale of stocks began, it became clear just how much interest the public had in this development. Within a short time all ten million shares, worth $200 million, were sold. Half of these were owned by the communications companies. The value of these stocks increased by leaps and bounds: originally issued at $20. per share, they soon climbed to $70. per share. It was now big business. Space exploration had proven effective not just to a few scientists, but also for the solution of real problems on the earth.

Without wasting any time, COMSAT began talks with other countries to create an international communications satellite organization. This was not an easy task; it became clear that the Americans, as well as the other nations, had their own interests, and each wanted its specific interests taken into account. In addition, it was unclear whether COMSAT was supposed to deal with the governments, or with representatives of the communications industry in each country. After a great deal of work the first treaties were finally hammered out, and on August 20, 1964 fourteen countries created the International Communications Consortium (INTELSAT). INTELSAT developed so quickly that just one year later forty five countries belonged to it. Today this number has almost doubled – eighty three countries now belong to the consortium.

This rapid growth had been anticipated from the very beginning. INTELSAT was conceived as an interim solution. After five years a new mode of government for the consortium was to go into effect. Until that time (and this later turned out to be a great mistake) power was divided within INTELSAT in such a way that the United States could never be outvoted.

The basic idea was simple. By 1968 INTELSAT was supposed to have at its disposal a complete system of three satellites in synchronous orbit. It was easy to estimate what percentage of the system's total capacity each country would use.

This provided an excellent formula for determining the ownership of INTELSAT. The decision-making power of the countries composing INTELSAT was figured on the same basis.

At first glance this key to the division of ownership seems reasonable. Let us not delude ourselves, however. It had no other purpose than to cement the U.S. position of power. Because of the great demand for information in highly civilized countries, America's portion of INTELSAT would never sink below 50%. In addition, COMSAT was made manager of INTELSAT, and this strengthened the position of the United States even more. This meant that almost all contracts for development went to American industry, thereby increasing its monopoly.

On the other hand, the developing nations were forced to help bear the cost of this system without ever having a voice in INTELSAT's decisions. The Soviet Union would have been in a similar position, if it had joined INTELSAT, since its communications needs are less than those of Switzerland. Consequently, the Soviet Union had to build its own satellite communications network which the countries of the eastern bloc joined. INTELSAT split the world into two camps – an unfortunate consequence which could have been avoided. Fortunately there have recently been signs that this can all be changed for the good.

Telephone Conversations Become Cheaper

Even if many countries disliked the Americans' total control over the first completely workable communications satellite system, nonetheless they had to admit that there were seldom failures. Even the launching of Early Bird on April 6, 1965 was a complete success. Soon after it had achieved its orbit over the Atlantic, the first tests showed that all of Early Bird's equipment functioned flawlessly. Television broadcasts, as well as telephone conversations, sent over 200 different circuits, went without a hitch.

On May 2, 1965 the satellite had its public premiere. About 300 million people in North America, as well as the countries of Eurovision, watched a transatlantic television program that seemed unsurpassable in its brilliance. The program was as clear as if it were being broadcast by a local station. Even the choice of broadcasts was a success. This was an impressive demonstration of how satellites could open up completely new dimensions.

The newscast that formed the beginning of the program exceeded anything ever seen before. Time and place became meaningless when pictures of the Nobel Prize winner Martin Luther King as well as others appeared on the television screen. The programs showed the two participating continents exactly as they were at the moment of broadcast. There was no longer any delay between the time an event took place and the time it was reported. The viewer was there in person. A new era of news reporting had begun, finally breaking down nature's last barriers.

The next portions of the program were just as historic. First came the broadcast of a heart operation from Houston. While this was very instructive for the surgeons watching, it was, unfortunately, completely incomprehensible to the layman. Next there were a few minutes of suspense as the FBI, Scotland Yard and the Royal Canadian Mounted Police allowed each other a look into their voluminous criminal files. This had unexpected results. A few days later a viewer in Florida identified a gangster wanted by the Canadians, and was able to tell the authorities the criminal's hiding place – all because of a satellite broadcast.

This incident shows quite clearly the various ways communications satellites can be used. It was purely coincidence that the first transmission of criminals' pictures via satellite led to the capture of the man sought. Nonetheless it has long been a common practice in America to exchange such pictures, fingerprints, and other data which the police consider important, between large cities via radio. This has met with great success. With the help of satellites, they will be able to expand this system in the not too distant future, so that even the most remote village will be included.

Just three weeks later another use for communications satellites was discovered. On May 24 at Sotheby's Auction House in London a very unusual auction took place. Television cameras scanned every object up for auction, not for posterity, but rather for a group of American auctioneers who were sitting in the New York branch of the auction house Parke Bernet, bidding along. It was the first international art auction in the world!

Although all these small accomplishments were very interesting, they could not divert attention from the fact that Early Bird had been launched for a completely different purpose. Its main function was to supplement the transatlantic telephone cables in their routine work. That is what it now did. Early Bird was turned over to INTELSAT on June 28, 1965. This was a step into the future. At a single blow the number of telephone circuits between the United States and Europe increased by 75% from the 317 already existing. In Germany this meant that, instead of forty eight circuits, there would now be sixty available.

It did not take long before the satellite was completely overloaded – so overloaded in fact, that protests were made. Early Bird could relay either one television program or 240 telephone calls; whenever a television program was being broadcast, a backlog in transatlantic telephoning developed. In addition, telephone charges were naturally determined according to the rent which had to be paid for broadcasting a television program. Thus, within six months $2.5 million flowed into INTELSAT's cash registers. A rough estimate showed that this was far more than was necessary to run the satellite profitably. For this reason, with the agreement of the Federal Communications Commission, the rates for all transmissions, except television broadcasts, were lowered at the end of 1966.

It is interesting that a branch of the American government exerts such an influence over international communications. This is not in the least beneficial to INTELSAT. Consequently, many countries, especially in Africa and Asia, have begun to reconsider their position within the communications consortium. Was this really the ideal solution? Should COMSAT, as the manager of an international union, need the approval of the American government before it could make any decisions – especially when these decisions affected not only the United States, but the entire world? That could not be right.

Indeed the Americans vigorously exercised the power they possessed within INTELSAT, almost to the point of extortion. They simply suppressed tendencies which they did not particularly like. In this case, however, all of the member nations pulled together. They all considered lowering the rates a good idea. An extensive analysis of COMSAT had shown that this could be done without any particular risk. The need for communications satellites would increase so much by 1971 that COMSAT's shareholders would still get a respectable dividend.

In anticipation of this increased demand, more and more powerful satellites were developed. Early Bird, or INTELSAT I as it is usually called, soon burst its seams. At 83.5 pounds it was actually a dwarf. INTELSAT II, the consortium's second generation satellite, had an effective weight of 190 pounds. With INTELSAT III, the third generation, the weight had already risen to 320 pounds, and the present satellite giants have the enormous weight of about 1,550 pounds.

The satellites' capacity grew with their weight. INTELSAT I, at a radiated power of 20 watts, was able to broadcast either one television program or 240 telephone conversations. INTELSAT IV satellites, by comparison, can cope with 12 television broadcasts simultaneously, or about 2,000 telephone connections. This is tremendous progress! In the short decade since the launching of INTELSAT's first satellite, the

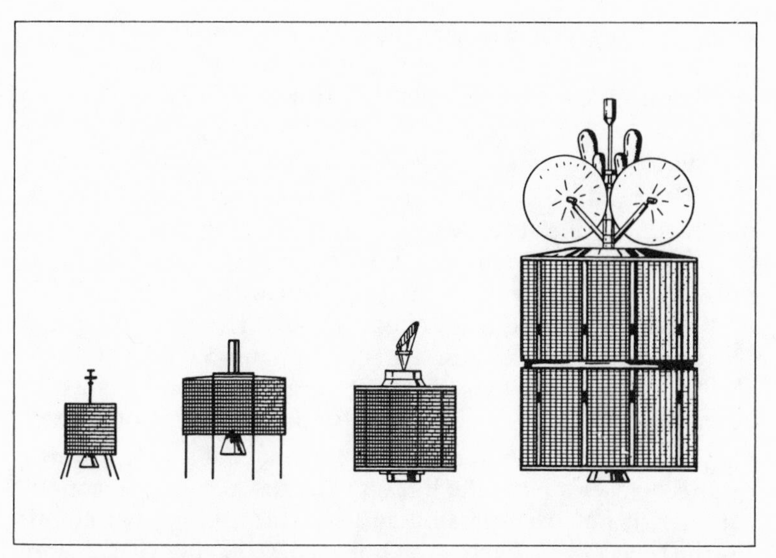

Comparative size of Intelsat satellites. From left to right, Intelsat I (Early Bird), Intelsat II, III and IV.

number of transatlantic telephone circuits has grown from 317 to 9,510, and the end is still not in sight. Every year the need for communications over the Atlantic grows by 35%, so that before long even the INTELSAT IV satellites will be unable to cope with the flood of information exchanged between the continents. This will signal the beginning of the INTELSAT V satellites, which are even more gigantic. These satellites will be the size of a three- or four-story building, in comparison with the INTELSAT IV satellites, which are about two stories tall. They will be able to relay neither more nor less than 48 television programs or 50,000 telephone conversations! This is the beginning of a new era.

Who would ever have dreamt this when cables gave us the first clear telephone contact between the continents? The usefulness of cables was only short-lived. Scientists are presently expending great effort in the field of modulation technology in an attempt to send simultaneous conversations over one circuit. This is being done in the hope that cables will not totally disappear from the picture. However, they still need to prove themselves. Every time cable technology advances by one step, satellites have already advanced two steps.

The result is a rapid decrease in the cost of relaying communications. For transatlantic traffic, using INTELSAT I cost $20,000 per circuit per year. With INTELSAT II satellites, the charges sank to $10,000, and with INTELSAT III they were already down to about $3,000. The rates for using INTELSAT IV satellites are only about $700, and with INTELSAT V satellites the costs are supposed to sink to about $30.00. This is an extremely impressive array of figures.

The decrease in charges is so phenomenal that it is worth taking a closer look at the figures. Thanks to communications satellites it has been possible to lower the charges per circuit for the New York – Paris stretch from $20,000 per year, before satellites, to $11,350 at the beginning of 1971. At the same time, the charges for the stretch from San Francisco to the Phillipines were lowered from $25,000 to $18,900, and from San Francisco to Honolulu from $17,000 to $6,700!

Of course not all transatlantic conversations are becoming cheaper at the same rate. Nonetheless the user does share in the savings. Thus, telephone charges for domestic calls have risen recently, while calls across the Atlantic cost less all the time. John A. Johnson, Vice President of COMSAT, announced the most recent decrease in rates with the introduction of INTELSAT IV satellites in 1971.

It became clear in 1969 just what effects such rate decreases would have. In February of that year the rates for exchanging television programs on the international level sank some 40%, due to the advancing communications satellite technology. It was the old game of supply and demand at work. Opening more circuits increased the demand for them, and within a short time the exchange of television programs between continents increased by 100%. This same effect and countereffect can also be seen in worldwide telephone communications. Since it has become possible to dial directly from some places in the United States to population centers in Germany at a reasonable cost, this is increasingly common. Long distance oral contracts or agreements are made, world commerce is expanding, and, consequently, the economy of the entire world is benefiting.

Some people like to brand space exploration as the expensive hobby of a few rich nations. For that reason, I have made these attempts to express in figures the savings which space exploration has brought to mankind. Perhaps this will justify it, and motivate new developments. My attempts must of necessity fail; they must remain only superficial. The figures for the actual usefulness of these communications satellites probably reach far into the billions. The problem is that the entire complex of results and secondary results is so extensive that it cannot be put down in absolute figures.

Worldwide Communications

Even though construction of the INTELSAT system progressed by gigantic steps, it was not completed overnight. In the beginning only America, Europe and North Africa profitted from the new technology. The first expansion was not made until after INTELSAT I had already outlived its projected life span. 1967 saw the launching of three satellites of the series INTELSAT II. They took their positions over the Atlantic and the Pacific. One was over the Atlantic, close to INTELSAT I which, to the technologist's great surprise, still worked satisfactorily after years. The other two were over the Pacific, so that the Far East was now also part of the regular communications network. There was still one hole over the Indian Ocean.

According to what was now an old tradition, this was the occasion for another public television broadcast. It took place on July 25, 1967, and had the theme "Our World". This time 24 countries on five continents took part. The number of viewers was estimated at 500 million. At an expense of $3.5 million a "super show" was presented, whose stage was the entire world: it involved lightning-fast jumps between all parts of the world, a gigantic feat of technology. Yet the appeal of something new had been lost. This was to be the last special demonstration of communications technology. From now on, everything became routine. After this even the most impressive color television relays from the most isolated regions of the world were no longer able to elicit such enthusiastic reactions as did those first technically poor broadcasts from the pioneer age of communications satellite technology.

This satellite network was finally completed on May 22, 1969 when the third satellite, type INTELSAT III, was launched and stationed over the Indian Ocean. Two satellites of this type had already replaced their predecessors over the Atlantic and the Pacific, so that now 90% of the earth's population was covered by an efficient communications satellite network. One of mankind's dreams was fulfilled.

There had been few failures along the way. One of these failures was the unsuccessful launch of the first satellite in the INTELSAT III series, and this had almost caused a great uproar. This satellite was intended to carry the broadcasts of the 1968 Olympic Games from Mexico to North America, South America, and Europe. A quick solution was needed. A new launch could not be readied in so short a time. However, problems are there to be solved, and a way out of this dilemma was finally found. It was quite original, but nonetheless it satisfied all parties. The solution came from NASA, which had often stepped in to help when the capacity of INTELSAT satellites had not been adequate.

When NASA had completed its experiments in the Syncom Program, it left the field of civilian active satellite communications entirely to INTELSAT. The satellites Syncom 2 and Syncom 3 wound up in the possession of the American Army. Gradually it became clear that INTELSAT was capable of building up a worldwide communications system via satellites. However, it was also clear, that in the first years INTELSAT would not be able to carry on the basic research which is so fundamental to future progress. The situation with regard to civilian satellite meteorology and navigation was similar: only NASA was in a position to carry on the necessary basic research. For that reason NASA decided to develop a new satellite program specifically concerned with general problems in applied satellite technology: the ATS Program (Applications Technology Satellites). With the help of stationary satellites NASA wanted to contribute its share towards making the fruits of space exploration available to people on the earth. The first ATS satellite was launched on December 7, 1966, and stationed over the Pacific. ATS 1 showed NASA's goals clearly. It served for communications, navigation, and meteorology equally. Its main tasks were to transmit telephone and television broadcasts between the continents bordering the Pacific, as well as to photograph the earth. These photographs provide meteorologists with a continuous picture of the weather over the Pacific, and are of enormous value in basic research. Anybody who has ever had the chance to see one of the excellent ATS movies, that show how the clouds gather together over the earth, will be able to verify this. (These movies are made by projecting a series of stills.) Occasionally ATS 1 also relayed radio transmissions between airplane crews and ground stations. It was truly a very versatile satellite.

On November 5th of the following year another satellite in the same series was successfully launched. At first it was stationed at latitude 95° W., not far from the Galapagos Islands. This satellite, ATS 3, was primarily intended to deliver additional pictures of weather patterns as they developed. ATS 3 suffered a fate similar to that of Syncom 2 a few years earlier. The scientists never left it in peace. The time of year was approaching when those clouds which lead to powerful storms would gather in the Atlantic and the Caribbean, and then strike the North American continent. Thus the satellite was shoved to longitude 47° W., because the meterologists wanted to avail themselves of this first opportunity to observe continuously what was taking place. The location over the mouth of the Amazon River was especially suited to this purpose. After having completed its task, ATS 3 returned to its old position. Shortly after that, however, it was moved again, this time to longitude 85° W., where it was needed for measuring the earth. All in all, this satellite led an extremely mobile life.

When it became clear that the INTELSAT satellites by themselves would not be able to broadcast the Olympic Games from Mexico to the Americas and Europe, it was this satellite, ATS 3, that NASA now placed at INTELSAT's disposal. However, since the capacity of ATS 3 alone was not sufficient, the old veteran Early Bird was also used. The problem was thus solved, although in a somewhat roundabout fashion. ATS 3 broadcast the video and the sound effects over the Atlantic. Additional commentary was sent by radio relay system to Canada, where it was then broadcast to Early Bird which completed the circuit. The Far East could follow the events in Mexico via one of the INTELSAT 2 satellites stationed over the Pacific. This time combined forces gave a great portion of mankind the pleasure of seeing the Olympic Games live on their television screens.

With the exception of a few isolated cases, INTELSAT was able to fulfill all expectations. In spite of that, the consortium was in upheaval. The colliding interests were too diverse. In addition, more and more countries had joined INTELSAT, and the time was ripe to establish INTELSAT's permanent governing body. On February 22, 1969 a meeting was held in Washington to take inventory and to readjust.

The Europeans wanted a greater emphasis on industrial use. The Third World nations did not agree with this. Why should they help to finance the demands of European industry? Japan and Australia were in favor of a quick readjustment, because they had profited much from the organization.

This was in opposition to Germany, France, and several other countries. They had used the system much less than expected, and did not want to lose their percentage of ownership, which had originally been set too high. Should COMSAT remain as manager? It had used part of INTELSAT's funds to establish its own laboratories, in order, as it said, to supervise industrial use. A majority of the member countries opposed this. The dilemma was complete. An agreement could not be reached, and the conference was ended. A year later it looked as if an agreement were possible, but appearances were deceiving. Again, differing interests collided, and the decision was postponed. Finally on May 21, 1971 a solution was found that everyone could accept. Although this was only an interim solution that was to last six years, INTELSAT's future was assured. The treaty came into effect on February 12, 1973. When Jamaica, as the fifty-fourth nation, ratified it on December 22, 1972, the necessary two-thirds majority had been attained.

As a result of this readjustment, American influence was severely curtailed. However, it could not be completely broken. For management, a general secretariat was placed beside the all-powerful COMSAT. By December 31, 1976 all of the member nations will elect a general director to completely take over COMSAT's duties. America's portion of the ownership also decreased. Its part shrank to 39.2%. Nonetheless it still owns much more than Great Britain and Ireland (10.8%), Japan (4.4%), France and Monaco (3.9%), and the Federal Republic of Germany (3.7%).

Perhaps this decision shows the beginning of a tendency which could one day culminate in the Soviet Union also joining INTELSAT. China's attitude probably depends primarily on Taiwan's membership in INTELSAT. Western observers have come to the conclusion that the Chinese would eventually be interested in joining the consortium. This is based on earlier agreements they had made with RCA concerning international communications connections. The Chinese had then let these agreements lay dormant, but they renewed them after their entry into the United Nations. Furthermore, they now possess three INTELSAT ground stations in Peking and Shanghai. The first of these stations they received as a present from the American government during Nixon's China trip in February, 1972. Nixon had brought them along on his trip so that the entire world could witness the historic stay of an American president in the People's Republic.

An INTELSAT ground station has also been erected in the vicinity of Moscow. Officially it serves as the terminal for the "hot line" between Washington and Moscow, which in the future is to run via satellite. However, the Soviet Union had much more equipment delivered than would be necessary for this purpose. This is puzzling, too, especially as Russia has also profited from INTELSAT in recent years. AT&T and Germany have set aside some of their INTELSAT circuits for connection with the Soviet Union. Could these be the first signs of renewed Russian interest in joining INTELSAT? Will it be possible to unite the world in the near future, at least in communications technology?

The Orbita System

Indeed there is something else that points this way. Russia's attempt to create an anti-INTELSAT has failed. The Soviet satellite communications system has never gone beyond the purely regional stage. The East bloc nations recognized what an important role INTELSAT played in international communications.

Soon after the launching of INTELSAT I the Soviets began their own attempts to create a satellite communications system. On April 23, 1965 their first test satellite Molniya 1A (Molniya means lightning) achieved an unusual orbit around the earth. It varied between 375 and 24,780 miles. Experts in the West concluded that the Soviets had planned to launch a synchronous satellite which then, for some reason or another, failed to reach its expected orbit. This presumption was too hasty, for that was not the case at all. This orbit had been chosen intentionally. It was most appropriate to conditions in far-flung Russia.

Generally, a system of three synchronous satellites is best for relaying communications throughout the world. With the simplest technology such a system blankets the area lying in a belt between latitude 70° N. and 70° S., where 90% of the earth's population lives. However, in the Soviet Union there are several inhabited areas north of this belt, and they would not be reached by a synchronous satellite system.

For ideological reasons the Soviets are opposed to having their people flooded with broadcasts from the "enemy". On the other hand, there is the question of the survival of their regime. They must be able to spread their own propaganda in the shortest and the quickest way possible to the very borders of their country. History has shown that enormous empires can endure only if they possess a good communications system.

Until late in the sixties the Soviets had to live with the fact that their northern and southern border areas retained a certain amount of autonomy in communications. The local television centers scattered all over these areas were only indirectly under Moscow's control. The capital was far away, and transporting television films took an enormous amount of time. Relaying broadcasts over thousands of miles was technically so vulnerable that, even though directional radio and cable were used experimentally, only a portion of the television programs sent in this way ever reached their destination.

Communications satellites offered a way out of this dilemma, as long as they fulfilled one basic condition: they also had to reach the northernmost parts of the country. The Molniya satellites did just that. In accordance with the laws of physics, a satellite moves faster the closer it is to the earth, and slower the further away it is. The Soviets took advantage of this fact in a clever way. They shot their communications satellites into an orbit which was strongly elliptical, and had an inclination of 65° at the equator. Thus it was possible for these satellites to reach both the northern and the southern regions of Russia. Now it was only important that the satellites be far from the earth whenever they were north of the equator. The Molniya communications satellites complete their orbit around the earth in twelve hours. By launching them in this way, the Russians guaranteed that these satellites would be over the Soviet Union for eight of every twelve hours. It was an ingenious solution.

Testing of their communications relay system lasted two years. During this time, the necessary ground stations were erected in Murmansk, Arkhangelsk, Irkutsk, Alma Ata, Novosibirsk, Ulan Ude, and Jakutsk. By the end of 1967 the Soviet Union was covered by a network of 25 stations. Meanwhile the number of stations has increased to more than 40. These and the Molniya satellites together form the Orbita communications system, the first regional satellite communications system in the world. We call this system regional, because its use is limited to one region, the Soviet Union.

Sites for the Orbita stations were chosen according to very strict criteria. They had to be equally spaced throughout the country, but, at the same time, they could not be more than seven miles from the nearest regional television station. Thus the Soviets managed to establish a functional system quickly: they utilized already existing television networks to the optimum. Every station was connected directly to the central in Moscow.

The launch of satellite Molniya 1F on October 3, 1967 officially inaugurated the Orbita network. Only six months later, development had progressed so far that television programs from the Moscow central could be broadcast to all parts of the Soviet Union. The country had grown a new nervous system.

The example of the Soviet Orbita network shows clearly what advantages a regional satellite system possesses when adapted to local conditions. This is especially important in large countries where certain areas are extremely difficult to reach. In fact, not even communications technology can penetrate to great expanses of some countries. Thus it is no accident that the Soviet Union was the first to implement many of the possibilities latent in modern communications satellites.

At the end of 1969 the Chief Engineer in the Office of Space Exploration of the Soviet Communications Ministry, W. Krylow, announced very proudly that his country had been the first to successfully broadcast a newspaper via satellite. The Soviets had found a new way of bringing the border regions of their country closer to the center of activity.

In the experiment that Krylow was referring to, a new means of contact had been opened between Moscow and Khabarovsk, a city in distant southeast Siberia. One page of *Pravda*, the central organ of the Communist Party, had been sent to the ground station in Khabarovsk via satellite. From there it was forwarded to the local presses. Transmission of such a page in large format only takes two to four minutes, so that it is possible to propagate news in far Siberia, almost at the same time as in Moscow.

By now this kind of broadcast has become routine in the Soviet Union. It has partially replaced photoelectric transmission via cable. Cables had already connected Moscow with twelve cities in Russia. The process did have the disadvantage that some of the telephone circuits to the eastern part of the country were blocked every day for a certain amount of time. Communications satellites eliminated this problem.

COMSAT also has plans for setting up a satellite newspaper service for the press. Although this service is still in the developmental stage at present, it is supposed to be ready for use in the not-too-distant future. Transmission will be in digital form, and ordinary facsimile machines will be adequate for reception.

There is likewise discussion about sending letters by satellite. This plan is not so far-fetched as it might sound. The main problem, preserving the secrecy of the letter, has been solved. In 1970, an electronic letter relay machine, developed by ITT, passed its general tests. Using this system, a letter could be sent from Washington to Chicago in five minutes.

In principle the process was very easy. The letter was put into a mailbox in Washington, where it was automatically opened and read. A computer scanned it and encoded the information into electronic signals, which were then broadcast to Chicago. The letter itself was destroyed. In Chicago the signals were decoded by a machine, and the original text was then printed out on paper. Finally, the paper was put into an envelope, sealed, and addressed, and the whole process was finished. During the entire time, no human being had seen the contents of the letter. For some time now ITT has maintained a test stretch between Chicago and Battle Creek, Michigan. The next step is to tie the communications satellites into the system.

How long will it be before the postal service is active in outer space? How long before it takes over telephone circuits and sends letters? How long before every citizen in the world gets his newspaper sent into his home, so that all he has to do is press a button in order to get *Pravda* in Hawaii, the *New York Times* in Vladivostok, the *Washington Post* in London, the *Daily Mirror* in Paris? Or maybe just the most interesting lead articles, or the sports sections, or political or cultural news? We are living in a time of upheaval, and it is quite possible that many of us will experience the culmination of this development in news dissemination which is both international and topical.

It is true that so far only the first tests have been made in this direction, if we exclude the Soviet newspaper broadcasts. Their broadcasts, however, are in no way intended to introduce a truly free spread of information.

Finally the Soviets undertook to expand their Orbita system beyond their borders. On November 15, 1971, Russia signed a treaty with eight other countries of the Communist bloc, founding an organization similar to INTELSAT. It was called Intersputnik. The eight countries involved were Bulgaria, the German Democratic Republic, Cuba, Mongolia, Poland, Rumania, Czechoslovakia, and Hungary.

Intersputnik remained an experiment. For one thing the Soviets exercised unchecked power in the organization (even stronger than the Americans in INTELSAT), and this did not contribute much to friendship among nations. In addition, it was obvious that no non-communist country was going to join Intersputnik. The competition from INTELSAT was far too great. For this and other reasons the organization has been unable to overcome its initial difficulties. Things have reached the point that two of the first member countries of Intersputnik, Hungary and Czechoslovakia, are now planning to join INTELSAT. Will the Soviet Union soon follow them?

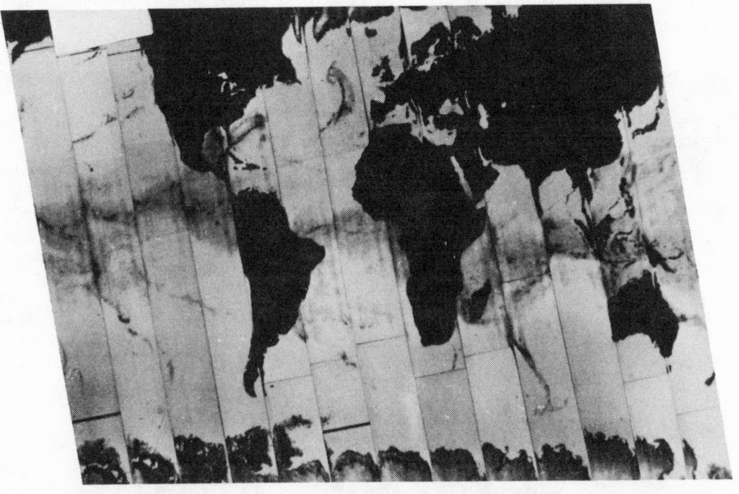

An unusual view of the earth. Nimbus 5's microwave radiometers penetrate the clouds effortlessly. This picture, taken on January 11, 1973, is made up of thirteen separate strips.

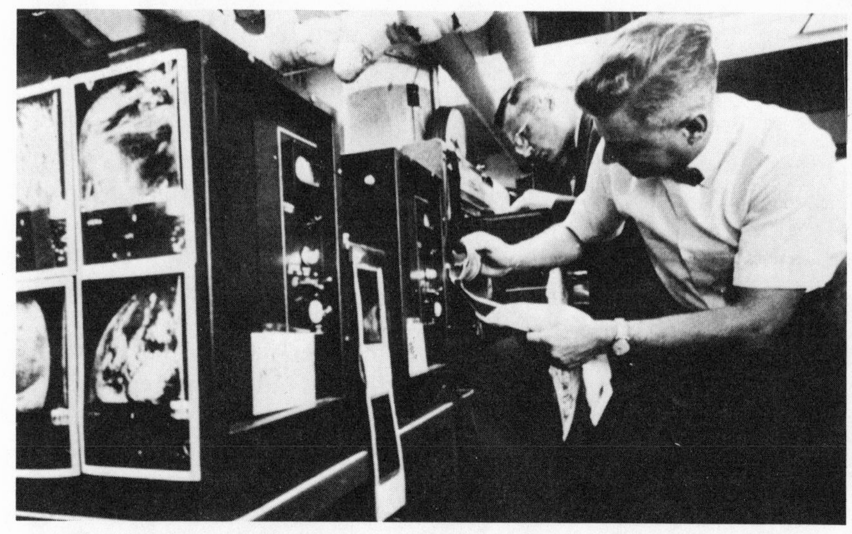

Suitland, Md. Evaluation of Tiros' photographs of the earth. Based on a detailed analysis of cloud structures (above), precise nephanalyses are made (below).

4. Education Via Satellites

Anik and Other Regional Satellites

"The white man has destroyed our peace. We hunted seal and elk before the white man came to our land. We had everything we needed to be happy, and we did not want anything else. Then the white man came to our land and brought trouble. He built enormous ears in the solitude to listen across the ocean. He dug holes in the earth, he built towers and looked for the black gold in our earth. The white man drove away the animals that we needed for our livelihood. In their place he build houses of wood, he brought us tobacco and fire water, things which we did not want, and he undermined the morale of our children. Once we were proud of our teeth, and now they rot just like the teeth of the white man. Our pride has been broken.

"Our children go to the white man's schools, and learn useless things. Stuffed full of what the white man considers culture and a basis for happiness, they come back, laze around our houses, and do not know what to do. They are torn between this new knowledge that they cannot apply and the old traditions that do not mean anything to them any more. Now they are unhappy.

"Hasn't the white man caused enough trouble already? Does he now have to send us a 'brother' who does not know our problems; a brother whom we do not trust, because we believe he is only going to hasten our downfall? We only ask the white man to leave us in peace, so that we can follow our ancestors. We do not need him."

We might wonder what Laurent Picard, president of the Canadian Broadcasting Corporation (CBC), and Andrew Cowan, director of the Northern Division of CBC, thought when at the end of November, 1972, a delegation from the Chiefs of the Eskimos and Indians of the High North complained to them with these or similar words. (Incidentally, this was the very first time that tribal chiefs had met with the heads of Canadian companies.) What had prompted this step was Anik, a Canadian regional satellite. It had been launched a few days earlier, and was mainly intended to open up the northern part of the country. Anik is the Eskimo word for "brother". Now those people who were supposed to get the most use from this satellite were protesting. This was an unexpected development which could not be handled easily.

Thus it is scarcely surprising that the delegation of chiefs sent to Ottawa by the newly founded Federation of Natives north of 60° returned home with their mission unaccomplished. A new meeting was agreed upon, but that was only indicative that it would be difficult to reach agreement.

This historical meeting in Ottawa reveals the whole misery of our times, indeed of civilization in general. We are proud of our awareness of our mission, but we forget entirely that cooperation among people is only possible if they understand each other. Maybe this meeting occurred at just the right time, so that those in positions of responsibility could learn from it, and know how best to employ modern technologies in the future – including communications satellites.

The problems in Canada which led to the launching of Anik, are similar to those in the Soviet Union. Of all the nations in the world, Canada has the second largest area. Vast regions of the country are swampy. Elsewhere thick forests make settlement difficult. In the north, temperatures during the winter regularly sink below –30° F., which is not conducive to colonization. Consequently, trade and industry are concentrated in the south, and the border with the United States is the most heavily populated area.

The north, however, has remained underdeveloped territory. Even the Northern Division of CBC could do nothing about that. For years it had made a commendable effort to supply topical film material to the completely isolated television stations located in the far north. Topical or up-to-date under these conditions means not more than two or three weeks old. Just transporting the films takes that long. Even telephone service suffers from the vagaries of nature. There are hardly any cables in the northern part of the country, so that the population must depend upon radio transmission. At best, this can be operated two hours a day, if at all.

Thus the developments in communications satellite technology seemed like a gift from heaven to the Canadian government. This is where the solution to the problem lay. They could only open up this land with satellites, at least in the area of communication. If they succeeded, one of the main hindrances to settlement of this area would be abolished. It seemed that rosy times were about to begin.

Through the action of Parliament, "Telesat Canada" was founded in 1969. Its mandate was to cover the country with a dense communications network. Working via a stationary satellite, it would connect even the most farflung regions with the population centers in the south. While the system was supposed to function on a commercial basis, its main goal was to open up the country.

From a purely technological point of view the planning of Telesat Canada was exemplary. In the northern part of the country two main centers – Frobisher Bay on the Atlantic coast and Resolute deep in the interior of the country – maintained television, radio and telephone contact with the large stations in the south – Toronto and Victoria, British Columbia. There were also a series of smaller stations. Depending on need, these could either receive CBC television programs or expand the country's radio and telephone network. For the most part, the stations that took over telephone service were located in areas with a population of less than 500.

It was the protests of those living in the far north, those who were supposed to profit most from the new system, that finally drew attention to this undertaking. During the planning stages no one had asked their advice, and the original aim of the project was gradually forgotten. The protests were justified. What kind of television programs did the CBC have to offer? They were nothing but worthless westerns and cops and robbers films, interrupted by endless commercials – for example, for toothpaste! This is the way television is in America. This was the new culture. Who was that supposed to help?

Anik might have had a chance to make a significant contribution by opening up and civilizing the country, if at the time equipment was installed some provision had been made to produce new television films aimed at the problems of the population. The "hardware" should have been supplemented by the necessary "software". This point was passed over silently. However, it is not yet too late. The government can still give the necessary impetus, and by the end of 1976 Canada could be the example of a nation that really does something for its people.

It is questionable whether the Canadian government will take this step. Since Anik started regular service on January 1, 1973, the bureaucrats seem to be satisfied. Telesat Canada's bankroll has grown. Money talks. Even before the launching, most of the ten available television circuits or equivalent telephone circuits had been rented. This satellite was developed on the drawing boards of the Hughes Aircraft Corporation, from the original INTELSAT IV satellite. The circuits were rented at a price of $2.5 million per circuit per year. At that rate, for two satellites in use over five years, it meant a pure profit of $200 million – a very lucrative enterprise.

These figures caused the calculating managers of American communications firms to take an interest in the project. Suddenly the old pioneer spirit awakened in these bosses. Money – one could talk about that. Why should only Canadians earn money from North American regional satellites? Why shouldn't the Americans seize the initiative themselves and buy satellites, if these were able to decrease large capital outlays, and especially if they could net a profit?

Western Union Telegraph Company started the procession. At the time it was in a hopeless situation. Its communications cables, connecting the west coast of the U.S. with the east coast, were completely overloaded. Twenty percent of the telephone connections between individual cities had to be carried over cables that belonged to AT&T. This cost Western Union $20 million per year. It was a ridiculous predicament. Western Union served the customers, and AT&T got all the profit. There was only one solution, and that was to buy satellites. The company drew the necessary conclusions. It ordered three Anik type satellites from the Hughes Aircraft Corporation, and simultaneously started building its own ground stations. After New York, Chicago, Atlanta, Los Angeles, and Portland, a ground station was to be established in Honolulu.

The company's risky position can be seen from the fact that it had not even obtained approval from the FCC for operating its own communications satellite. Ordering the satellites could easily have turned out to be a misguided investment. Eventually the situation cleared up. At the beginning of 1973 the FCC allowed the establishment of regional communications satellite systems in the United States.

The omnipotent influence of the FCC can be felt worldwide, and shows itself again and again. For example, until now, no country has succeeded in convincing the Americans to launch a communications satellite to compete with INTELSAT. The Canadians had to promise to use Anik only for regional broadcasts. Japan, meanwhile, is developing a new booster rocket. By 1980 it is supposed to have reached the point where it will be able to launch a Japanese communications satellite in a synchronous orbit. To a certain extent Europe too has made itself independent, by its decision, reached after long debate, to support the development of a

French rocket, capable of orbiting a satellite. This is especially important because nobody can say with assurance that the Americans would make a rocket available for launching a European communications satellite. This would be a satellite especially intended for relaying communications in a north-south direction, for which the INTELSAT satellites are unsuited.

In spite of the risk, Western Union did not long remain the only communications company flirting with the idea of owning its own relay satellites. The free market encourages bitter competition; it is not possible for one company to be at the top of the ladder without other companies immediately trying to climb up. News of Western Union's plans soon spread, and in a very short time other firms also announced their interest.

At the end of 1974 the American Satellite Corporation planned to launch its own communications satellite: one of the Anik type, ordered from the Hughes Aircraft Corporation with ground stations to be built in Los Angeles, Dallas, Chicago, and New York. This will mean competition for Western Union, but there is no danger that both firms will not be able to profit. The need for communications circuits is great enough to accommodate more than one system at the same time, in spite of their enormous capacities.

Other companies have even gone one step further in their planning. For them the Anik satellites are not large enough. They are waiting for the next generation of satellites, due soon, so that they can profit even more from the new technology. AT&T as well as RCA Globcom/Alascom have already expressed this idea. They want to include Alaska, Hawaii and Puerto Rico in their network.

Under these conditions it is no wonder that the managers of Hughes Aircraft Corporation are rubbing their hands. This is big business. Something very unusual has begun. If this trend continues, then the time is no longer far away when the purchasing agent for a large corporation will be able to order a communications satellite from a catalog. They might even be satellites which were mass produced, because of the great demand. Thus they would be cheaper, and would attract even more customers. The end of this spiral is not in sight. How will the world of tomorrow look?

Only a few companies can carry on such pioneer work. Does the TelePrompter Corporation belong to this group? This company has 140 cable television systems in the United States alone. For the annual congress of the National Cable Television Association (June 17–20, 1973) in Anaheim, California, they staged a demonstration designed to show the marriage between satellite television and cable television. Would this be successful? Perhaps. In any case it was good that the participants in the congress were present when a portable antenna, 25 feet across, relayed the first cable television program from the east coast via Canada's Anik. It originated at the ground station of the American Satellite Corporation in Germantown, Maryland. It was also good when, at the end of the congress, a truck left with the antenna to carry out a twelve to eighteen month good-will tour spreading this idea throughout the country. It was a promising beginning which could immensely increase the utilization of regional satellites.

Military Interests

One such development which we must not ignore, because it will inevitably lead to a revolution in the communications sector before the end of this decade, is the direct reception of television programs from satellites. Up until now television contact over the ocean could only be carried out by using large ground stations. Since the incoming signals are so weak at the receiver, they could only be separated from the static with a great deal of effort. The Americans have come a step closer to solving this problem. Thanks to the use of new energy sources in outer space, satellites are able to broadcast at a much higher output. New satellite antennae also help strengthen broadcasts.

It has been possible to receive voice broadcasts from satellites for a long time using only small antenna. For one thing, it is much easier to transmit small amounts of information. Secondly, these broadcasts have special importance for the military, who have always been a motivating force for science. The real reason they are interested in this technology is that it would provide a worldwide nervous system, including even the smallest military unit, the individual soldier in the last row. Whether he is in the jungles of Vietnam, in the deserts of Africa, or in the swamps of the Amazon, he would only need to press the button on his portable radio, his walkie-talkie, to contact headquarters.

This development evolved without publicity during the sixties. It is closely connected with the satellites called LES (Lincoln Experimental Satellite), the experimental satellites of the Lincoln Laboratories at MIT. They are primarily intended for experiments in communications relaying at the fringes of the radio wave spectrum, and have accomplished astonishing things.

In 1968 a remarkable experiment was undertaken with LES 5. The purpose of the experiment was to determine whether it is possible to set up communications between ships, air planes and landbound earth stations at relatively small expense and effort. It proved to be feasible. However, since ships and airplanes are fairly large, they could, if necessary, carry larger antennae than foot soldiers, whose mobility must not be restricted. This led to the next development.

In 1969 Electronic Communications Incorporated introduced a new walkie-talkie. It was no larger than the one the military was already using, and liked so well. The regular military walkie-talkies had a range of several miles and were so light that every soldier could carry one. They were extremely well suited for communication with a company or a battalion.

The new walkie-talkie from Electronic Communications, including batteries and antenna, weighted seven pounds. This is actually nothing remarkable in itself. What made it different from other radios was the fact that its range was worldwide. It operated via satellites. With this instrument any foot soldier, any pilot who was shot down, anybody whose ship was wrecked could contact his comrades over the greatest distances. It was finally successfully tested with the satellite LES 6.

Because of the favorable outcome of the LES experiments, it is now possible to develop a new nervous system for the American military. On February 9, 1969 the first 1600 pound tactical communications satellite was launched into a synchronous orbit around the earth. In 1975 or 1976 an operational network of four synchronous satellites will follow this first satellite. In actual point of fact America's future could very well depend on these satellites. A portion of their circuitry is reserved for communications relays of the utmost urgency: for the president's headquarters, for warning and monitoring systems, some of which are always in the air even in peace time, and finally for the bombers of the Strategic Air Command, which at a moment's notice can reply to any attack with its own atomic attack.

The creation of another satellite system is also planned. This will bring the SAC bombers stationed in the polar regions into the system. To achieve this, the satellites will probably be sent into elliptic orbits, similar to those of the Soviet Molniya satellites.

We can scarcely estimate what the damage would be to the Americans if this satellite were made inoperative at a critical moment. Who can blame the American military for the fact that they were originally unable to accept synchronous satellites, since these satellites were so exposed, and thus extremely vulnerable. A system of randomly scattered communications satellites was much more secure. The first strategic communications system of the United States, the IDSCS (Initial Defense Satellite Communications System), began operating in 1966, and consisted of satellites which slowly meandered over the equator.

The original advantages of such a random system have now been lost. Today all satellites are equally threatened by destruction. Combat satellites and orbital weapons have been fully developed, so that an outer space where vehicles fly only for peaceful purposes has become an illusion. This was made very clear on January 29, 1967, a few days after the Soviets and the Americans had signed a treaty which prohibited the orbiting of missiles for mass destruction. However, they did not at the same time ban development of "killer satellites" and atomic weapons that would stay in space only a very short time.

On this day, January 29, the Soviets tested a new weapon which entered the history of the cold war under the initials FOBS (Fractional Orbital Bombardment System). We have long known that its purpose is to transport atomic bombs to any desired target on the earth. Thus it combines the properties of both rockets and satellites. Because of its enormous speed, it can reach an orbit around the earth within only a few minutes. Once there, it does not remain, but, suddenly braked, falls like an eagle straight towards its target. In this way it infiltrates the classical early warning system of the western world. Unlike intercontinental rockets which, because of their low speed, travel thousands of miles into space in order to reach their target, the FOBS fly only for a short moment above the horizon of the radar installations.

A strange event occurred in the summer of 1970. Farmers in Kansas, Oklahoma, and Texas could not believe their eyes when they saw glowing tracks of light cross the skies. As it turned out, these were pieces of the Soviet satellite Cosmos 316 which had burnt up over the United States. Several of these fragments reached the earth's surface where they were collected and examined by military specialists.

The results of these examinations were astonishing. The

recovered metal plates were so thick and solid that the specialists immediately assumed Cosmos 316 had been a bomber. Was it only chance that the satellite had followed the typical FOBS orbit? Apparently what they were dealing with here was a secret weapon. Many of these weapons still circle the earth only because attempts to bring them back down have been unsuccessful.

Long before this discovery the Americans had signed a treaty which obliged them to return to the country of manufacture any "driftwood" from outer space. For this reason they returned the pieces to Moscow. Here the Americans had their second surprise – the Soviets acted dumb. They pretended they knew nothing, and they denied emphatically that the fragments came from one of their satellites. Today the pieces are back in Washington, where they are kept under heavy guard.

In addition to FOBS, the Soviets developed a second weapons system, this one aimed specifically at destroying satellites. Western observers reported the first tests in February of 1971. The satellite Cosmos 394 was sent into an orbit 375 miles above the earth. A few days later Cosmos 397 was launched. It slowly approached the first satellite, and then, with a single blow, destroyed it without a trace. The wreckage was clearly visible on radar screens. Similar experiments have been reported over and over, and today these "killer satellites" are fully developed.

The time is past when a satellite in a nonsynchronous orbit could be considered invulnerable. Additional combat systems are already being tested. Some of these destroy the essential components of a satellite directly from the earth. The satellite itself remains in space as so much junk. The magic word behind this system is "laser".

Who does not think of death rays when he hears the word laser? Even in most recent years such an idea seemed like something out of science fiction, but, appearances were deceiving. In 1968 at the Air Force Special Weapons Laboratory, Kirtland Air Force Base, Albuquerque, New Mexico, Project "Eighth Card" was begun in the strictest secrecy. Probably only the atomic bomb project during the Second World War was as confidential. In June 1971 the test area was expanded further. Then strange things began to happen, Several times every day a mysterious "cannon" was fired up, and at the very same instant a pile of wood two miles away burst into flames. Was this a miracle? No, the laser cannon functioned at the speed of light! Velocity is not a matter of witchcraft. When these cannon are fully developed, they will only need to be aimed at a satellite in order to instantaneously destroy its delicate sensors and antennae, leaving no possibility for defense. A dream has become reality. In spite of this, the United States continues to work on new ways to protect their military communications satellites. In the future special sensors will signal the approach of any enemy satellite. The synchronous satellites will be outfitted with equipment enabling them to make the tiniest course corrections automatically, without any command from earth. Such corrections are necessary now and then, because of the small positional changes caused by the extremely thin atmosphere and the pressure of solar radiation at 22,375 miles. This equipment has already been tested successfully with LES-type satellites.

Furthermore, research is being directed toward eliminating endangered ground stations in foreign countries. The satellites will be able to establish direct contact with each other without making a detour over the earth. Independence from the earth is the motto for the seventies. This is yet another way for safer communications, even though, as we have already stressed, complete security has become an illusion.

Catchword: Direct Television Reception

In the meantime, the civilian communications technologists are mastering the problem of direct television reception from a satellite. Without going into details, we can outline the basic ideas responsible for the first successes. At the beginning of synchronous satellite technology, the weight of the communications satellites was less than 220 pounds. This meant that the sender and receiver were both fairly small. The output of the equipment was limited by the capacity of its power source. For the most part, power was generated by converting solar energy into electrical current. This resulted in limited performance. Consequently, enormous receiving antennae were necessary on the earth to separate the satellite's incoming signals from background noise. For example, the first receiving antennae at the German ground station in Raisting in Upper Bavaria weighed 308 tons!

Such a system is naturally unprofitable, since the number of ground stations far exceeds the number of satellites in use. There is only one way out of this dilemma: the power output of the satellite has to be increased, and the quality of its broadcasts has to be improved by using larger antennae. In this way, the ground station's efforts can be reduced immensely. The result will be that the antennae now sold in stores will be adequate to receive television broadcasts directly from a satellite.

It was many years before development along these lines had progressed to the point where the results can now be enjoyed. The experimental satellite ATS F will be launched soon. Not only is ATS F a technological marvel, it is also a daring attempt to lead the world into the third millennium. In recent years a wide variety of people has stressed emphatically that the world is rushing towards a catastrophe. The earth's natural resources are almost depleted, hunger is spreading faster and faster, and the population explosion continues. Is this an inevitable progression into the abyss?

We can already see the first indications of a new world in the most isolated areas of the United States: in the lonely mountain villages of the impassable Rocky Mountains, almost hermetically sealed off from the surrounding world, and in Alaska, which is so inimical to man. Antennae and television sets are being installed in schools and in municipal offices, in kindergartens and in other public buildings. Is this normal procedure? Basically yes, but on a closer look, we notice one thing: these television stations are not meant for regular programs with all their westerns and advertisements, but rather for programs designed to help the village farmers, and for programs in health education. These programs are all broadcast centrally from the new satellite ATS F—satellite communications serving mankind!

Using satellites such as ATS F for developing countries opens up entirely new perspectives for the future. India is the first country outside of America that has recognized the possibilities. When NASA has finished the experiments it wants to do with ATS F, it will send the satellite along the equator to a position over the Indian Ocean. Here a new adventure in the exploration of space will take place: the education of an entire nation.

Since her situation is precarious, India has realized how valuable a satellite can be. Her political leaders have long known that they will only be able to bring their people to a respected position in the world by educating them and eliminating illiteracy. (70% of the Indians are illiterate.) Up until now, however, all their attempts have met with failure. Only a small middle class has profited from the federal projects. The broad masses are just as uneducated as ever.

The following is a typical story. Not too long ago, through Aid to Developing Nations, several Indian villages got a fairly large shipment of birth control pills in order to combat overpopulation. The villagers, however, did not know what they were supposed to do with these small round things, so they fed them to their chickens, which promptly laid no more eggs. This is precisely what we want to avoid in the future. Whether the story is true or not makes no difference. At any rate, it certainly hits the nail on the head: this is India's plight.

How can it escape from this situation? The problem

might not be quite so difficult, if the population were not so widely scattered. For the most part India's population is poverty stricken, the average yearly income is about $130. On the entire subcontinent there are 560,000 villages, and only 2,700 cities. The Indians are a people that belong to innumerable tribes, speak different languages, and are members of different religions and cultures. It is a colorful mixture. In a nation that has 14 main languages and 200 dialects, where do we start with education? How are we to stem the tide of the population explosion? How are we to increase food production?

With all this in mind, the Indian government has been flirting with the idea of using regional television stations to solve the problems of education. They envision education for everyone. The stations would broadcast the programs in the language of the local population, taking their culture and religion into account. The benefits would be enormous. Unfortunately, that is also true of the costs, and it would be a long time before the stations could be set up, much too long!

Thus the Indian space exploration organization, INSAT, (Indian National Satellite) suggested borrowing ATS F from NASA. It was only in this way that they could make immediate broadcasts which might realistically promise success. The satellite is ready. The large number of television sets which would have to be set up in the schools and in the government buildings of the many villages could be produced in a relatively short time by Indian firms. This would also give an impetus to national industry. The only things missing then would be a broadcast station and the necessary programs.

This is a crucial point indeed. What good is such an educational system without video materials? Lack of such materials would mean that remote Indian villages would also receive American westerns and cops and robbers films. The same situation has already happened to Indian radio: the stations of All India broadcast almost nothing but light music, to the great joy of the teenagers.

The Indian government has decided to pursue the possibilities of education via satellite, despite its drawbacks. One of the problems, for example, is that the three television stations are only broadcasting in a few languages, and can reach only a small portion of the population. Furthermore, the cultural and religious requirements of the people can only be taken into account to a very small degree.

The educational programs for adults and children alike are produced in India itself and are broadcast by a station in Ahmedabad, north of Bombay. Their offerings are very extensive. Seven hours of education are broadcast daily: four hours of reading and writing, two hours of agricultural advice, and one hour on the use of birth control pills, or, as this subject is officially called, family planning. Only 5,000 Indian villages will be equipped with television sets to start with. If the system works, it will be expanded to all 560,000 villages.

Will this monumental experiment to free India from ignorance, hunger and disease succeed? Many people have grave doubts. Critics think India must first solve her other problems. Road construction is in a very bad state, and only 71,000 of these 560,000 villages are electrified. How will they service and repair the television sets regularly? How will they deliver the gasoline necessary for the project?

These counter-arguments are not just fabrication. India's radio service has long been faced with similar problems. A few years ago an investigation was made. A commission inspected the radios which All India Radio had sent to the villages. It came to a very depressing conclusion: approximately 50% of them were no longer in working order! Television sets are even more apt to break down. Will this bring about the collapse of the project? Certainly no one hopes so. The entire world wishes India success.

If it is possible to receive television broadcasts from the ATS F satellites with small antennae (about ten feet across), then surely it will be possible to receive television broadcasts from satellites on home television sets, with no special equipment. This idea scares the Soviets. Can the Communist ideology survive this, or will it mean the gradual "decline" of the enormous empire run by the successors to Lenin and Stalin? The leaders in the Kremlin suddenly face a dangerous development which is not to their liking.

On October 12, 1972 the Soviet Secretary of State, Andrei Gromyko, made a motion before the United Nations General Assembly. By introducing some control over satellite broadcasts, the Soviets hoped to alter the course of this development before it went too far. They demanded that broadcasts receivable on home television sets be restricted, if the government in the receiving country objected to them. It is a question of borders. How can one limit the area of reception from a stationary satellite that covers one-third of the earth? Wasn't George Bush, the head of the U.N. delegation from America, right when he emphasized the fact that such censorship would endanger the most basic principles of the free world? Bush must also have had the consequences of such censorship in mind. If the restrictions were disregarded, the country where the program was received would have the right to destroy the satellite, even at the slightest provocation. According to Gromyko, all the following programs would be forbidden: any which disturbed international peace, those affecting domestic affairs, those restricting fundamental human rights, those dealing with violence, pornography or drugs, those directed against the culture, customs or traditions of the receiver, those giving false information, those dealing with war or racial problems. It is well known how liberal the Soviets are in the interpretation of these individual points. Anything "repulsive" comes from capitalist countries. Must these disputes be carried on in outer space too?

At the beginning of 1973, this motion to censor satellite broadcasts for direct reception was passed with 102 votes in favor, against 7 abstentions. A United Nations commission is now struggling with the task of working out the exact international regulations. This is a tragic failure of the intellect.

Once more man, with his petty medieval quarrelling has succeeded in blocking the direct path to the future. Even the best technology is not worth anything here. Space exploration provides us with the means for solving our problems: how we use them is another matter.

The sky over Europe is seldom as clear as it was on August 13, 1973. The satellite NOAA 2 seized the opportunity to take this impressive picture, using a wide-angle lens.

5. Earth: The Unknown Planet

The World of the Geographers

May 4, 1964 was a day like any other. The sun stood high in the sky, and the streets were full of people. Yet it was different from other days. A feverish disquiet had seized France's geodetic surveyors, the men whose job is measuring the earth. On this day they were going to begin an unusual large-scale experiment at five different places in France and North Africa: Lacanau near Bordeaux, Agde on the French Mediterranean coast, Oletta on Corsica, Hammada du Guir in the western Sahara, and Ouargla in the eastern Sahara. In each of these locations the instruments were being inspected for the last time: precision cameras with shutters timed by precise instruments. As soon as the sun disappeared over the horizon, the cameras were aimed at the sky. Every night for a period of three weeks they planned to photograph the balloon satellite Echo 1 simultaneously from these five different places, before the background of fixed stars. This experiment was to demonstrate that, with the help of satellites, it was possible to represent all parts of the world on a common system of coordinates.

Before the launching of Echo 1 this had not been possible. This does not mean that all previous atlases were wrong. Their scale is much too large to show any errors in the measurement of the earth's size, but there are errors. For purposes of surveying, the continents are merely isolated islands on one large ocean, and they cannot be related to each other. The earth's oceans are obstacles for geodetic surveyors just as in communications technology, an obstacle which can only be overcome with the help of satellites. (It should be mentioned here that in the case of Europe and North Africa the obstacle, that is, the Mediterranean, is so small that it had already been overcome by classical methods of surveying.)

The reason is obvious. If the geographic coordinates of a place are determined astronomically, that is by observing the stars, then the instrument that is used to fix these coordinates must be straight up and down, that is, its axis must point at the earth's center of gravity. This is the catch. The earth's field of gravity shows irregularities, so that the perpendicular does not actually point to the center of gravity. For this reason, calculations do not agree. The variations are slight, but noticeable. Looking at a place whose coordinates have been determined astronomically, from another place whose coordinates are also known, we can determine the exact differences between them, and we can then juggle these data around until the differences are minimized. Through this "compensation" we can attain a practical reference system. We can then figure out the geographic coordinates for every place in a country.

This process of compensation must fail on a global scale, since it is impossible to observe a place in Europe from America, for example. For this reason every part of the earth has its own system of coordinates, and the individual systems, when compared with each other, show errors of several hundred yards. How can the geographers solve this problem?

When the first satellite circled the earth in the fall of 1957, peeping away, the solution to the problem was in the air, so to speak. However, it was some time before the geodetic surveyors made use of the new tool that the satellite provided. There, high above the earth, they had an excellent, independent coordinate system. Satellites are so far from the earth that they can be observed at the same time from different continents, even on opposite sides of the ocean. If the observation is made by a minimum of three stations, then, from the measurements of two stations with known coordinates, the satellite's position can be calculated. It is then possible for the third station to determine its own coordinates. The necessary compensatory calculations would be made possible by using many stations throughout the world, and the result would be a worldwide system of coordinates.

This was the idea that the French geodetic surveyors put into practice in May, 1964. When they finished their observations on May 26th, they had a stack of photographs of Echo 1. These pictures, in the form of a cipher, were neither more nor less than the technological bridge between North Africa and Europe necessary for surveying. Soon the results were known. With the help of satellite observations, they determined the relationship of the two continents to each other within 35 feet. Geodetics had entered a new field of activity.

We can scarcely estimate the advantages this improvement in the methods of surveying the earth via satellite will bring in the future. After all, we no longer live in the time of Fernando Magellan. At the beginning of the 16th century, when he first sailed around the world, he marked down on his map islands in the South Seas. Because he had noted their position too imprecisely, these islands were only rediscovered decades later. Even today one has to reckon with errors of up to one and one half miles when trying to find isolated islands. In the future even that will be too much. Because our problems to procure more food and raw materials are steadily increasing, there will be no alternative: we will have to make use of the treasures in the sea. There, where all orientation points are missing, the most precise surveying methods are essential.

Echo 1 was not planned as a geodetic satellite. Its main function was to test communications relaying via satellite. Nonetheless it did exhibit characteristics which were of help to the geodetic surveyors. It traversed the earth at a great altitude, so that it could be observed simultaneously from widely separated places. It also shone so brightly in the night sky that it could be photographed against the background of fixed stars even with small cameras.

However, the surveyors were not completely satisfied. Precisely determining the location of a place presupposes an exact knowledge of that place at a definitely determined point in time. Since the speed of a satellite is several miles per second, the observations made by the various stations had to be synchronized to a thousandth of a second before they could provide valuable information! How were the technicians supposed to do this?

The geodetic surveyors employed a very simple principle in photographing Echo 1. During the actual photographing, they interrupted the exposure several times, so that on their photographic plates the satellite's track did not appear as one long unbroken line, but rather as a dotted line. The length of time that the exposure was interrupted was exactly predetermined. However, since not all of the observation stations could interrupt their photographs at precisely one thousandth of a second interval, extensive calculations were necessary before the photographs could be evaluated. Nonetheless, they obtained good results.

In spite of this success, the surveyors looked for even better methods. Finally one of the scientists hit upon a glorious idea: all they needed to do was equip a satellite with spotlights that could shine brightly enough to be visible from the earth. If these spotlights blinked regularly like a signal fire in the heavens, then the synchronization of the measurements was guaranteed from the start. In addition, the satellite itself could send out the precise time signal so necessary for calibration.

This was no sooner said than done. All those organizations interested in surveying the earth pooled their efforts to construct a new satellite. It now became apparent that the military wanted a share in this new technology. Their main interest was to study the earth in such detail that they could accurately steer an intercontinental missile to its target from any distance. In a total war, that would mean absolute destruction. For some this was a disaster, but for others it was the final goal unattainable up to then. To this end the military wanted to initiate basic research. Responsibilities were divided along traditional lines; besides NASA, all the armed forces were involved. The name of the satellite project, ANNA, contains the first letters of the names of the participants: Army, Navy, NASA, Air Force.

Soon afterwards they produced the first blueprints for the satellite. Its most remarkable characteristic was four glaringly bright Xenon lamps. Their illumination is so extraordinarily intense that they are often used in simulation chambers on the earth to imitate the sun's glaring light. In every short flash they developed the brightness of eight million candles, almost unimaginable! They were supposed to help science.

Unfortunately, not only science benefitted from them. The Pentagon was quite aware of the significance of the project. Were other scientists also to receive data which were important for the military? A short battle took place in the Department of Defense. Ignorance was victorious. The project was stamped "Secret", and NASA was unceremoniously booted out. This was the first clear evidence that civilian scientists could not count on the support of the military in the future, even though their interests were the same.

In the case of Project ANNA, however, the military had second thoughts. After long vacillation, NASA was taken back into the circle of participants, so that the satellite, launched on October 31, 1964 justly bore the name ANNA 1B (and not something like "ANA" 1B). The geodetic surveyors had their signal fire in outer space, but they were still not satisfied.

ANNA 1B did send regular flashes to the earth, but there were only twenty series of five flashes each daily. The energy supply on board the satellite was insufficient for any more. These Xenon lamps were pure energy guzzlers. This did not stop the military from building additional "signal fire" satellites, known by the name Geos (geodetic satellite).

At the suggestion of the American Army, ANNA 1B was also to be tested for geodetic surveying of the earth. When tracking satellites optically, the angles are measured to determine the coordinates of a distant place. The same results can be obtained by using measurements of distance. If the distance from one satellite to several stations is known, then the satellite's precise location can also be determined. Several successive measurements of the distance between another station and the satellite would then give the coordinates of that station.

The American Cubic Corporation developed the SECOR process based on this principle. One after another, four ground stations send out radio signals at an interval of 50 milliseconds. The satellite receives them, and broadcasts them back to earth. Technologists can then calculate the distance from the length of time it takes the signal to reach the satellite and return, or from the phase shift of the signal. Sending repeated signals is necessary if the satellite is to get by with just one receiver and one sender. This is where the name SECOR comes from. It is an abbreviation for Sequential Collation of Ranges.

The SECOR process was also tested on ANNA 1B. Again it was the military that carried out these tests. They built another type of satellite unofficially named SECOR, which gives us an indication of its activity. Later, these satellites were christened with the not very pretty name EGRS (Electronic and Geodetic Ranging Satellite).

Meanwhile the civilian scientists had to contend with insurmountable difficulties. The military did not want to let anyone else see their results, which were protected by the halo of national security, and had cut themselves off from the civilian scientists. These scientists needed an international network of observation stations, open to everybody, before they could create a worldwide geodetic system.

In the mid-sixties a solution finally appeared. On June 24, 1966 NASA shot another balloon satellite into orbit 2,600 miles above the earth. Pageos (Passive Geodetic Satellite) was to serve as the target for all those involved in surveying the earth. True, this was a return to the beginning, to the time of the Echo observations, but time had revealed that this method was not at all bad.

The American National Geodetic Survey now helped organize the ground stations. (Interestingly enough, it worked in cooperation with the Department of Defense. Apparently the military had to have their fingers in every pie.) Finally 45 stations were set up, scattered over the earth at distances varying from 2,500–2,800 miles. The network had only two gaps: one in the South Pacific, where no island at all could be found, and one in the Soviet Union. Even in science political considerations cannot be ignored.

The planning of the project was unimaginably complex. Every station received its own quartz clock. This way the simultaneity of observation was guaranteed. How were these clocks themselves to be calibrated? It was very simple: there was a technician for this purpose. He was to travel around the world with an even more precise atomic clock, and synchronize all the clocks. The time signals regularly broadcast over radio were not sufficient for calibration, since the amount of time it takes them to reach the stations depends on atmospheric conditions. A new profession had arisen, the profession of the "atomic-clock-world-traveler".

Finally the observations could begin. By November, 1970, the photographs were taken: a total of 3,600 pictures. Evaluating them took about three years. The geodetic surveyors can now breathe a sigh of relief. About one and a half decades after the launch of the first satellite, a worldwide geodetic system is ready and available to all scientists. This is a global system in which the location of 45 places on the earth is plotted on a uniform system of coordinates, with a margin of error of less than fifteen feet. Surveying the world has taken a giant step forward.

A Look in the Atlas

We have already mentioned the need to tap the ocean's almost inexhaustible sources of food and energy. Whatever progress is made in this direction will play an extraordinarily large role in mankind's future. Nonetheless we should be clear about one thing: the basic picture of our world is unchanged.

The photographs that the satellites have radioed back during more than a decade verify the fact that our earth was originally surveyed correctly. These photos show the same picture that all atlases do: an enormous ocean broken up by several continents. The continents, in turn, are composed of a number of smaller areas, whose characteristic outlines are familiar to everyone. Italy retains the form of a boot; Scandanavia still looks like a large dog in northern Europe. Over all of them lies a thick cloud covering which never conceals the entire world, but rather only a few countries or provinces at a time.

It is in the details that differences arise, pointing to the fact that there are still regions on the earth seldom visited by man – regions where nature is still untouched. Opening up these regions will be the task of the next decades. New centers of civilization will arise; new agricultural areas will develop to fight world hunger. We will return to these points later.

Naturally, the survey pictures from the satellites showing us large areas in a single view, are not completely identical with the maps in atlases. This would be impossible, since a sphere like the earth can never be copied exactly on a two dimensional surface. Distortion must be expected here and there, and the amount depends to a large extent on the way the earth is projected onto a plane. The larger the section that is projected, the more noticeable is the distortion. For the specialist this is nothing new. The layman, however, might not be aware of this fact. Thus, recent pictures from the American weather satellite, Nimbus 5, could cause hopeless confusion;

these are composite pictures made from individual strips. For the first time they show the entire world without the disturbing cloud fields which in reality always cover our planet. This sensational composite was only possible because Nimbus 5 was equipped with microwave sensors. These sensors register the natural radio emissions of the earth, and, since radio waves pass through clouds without any problem, unlike visible light, cloud-free "pictures" of the earth could be made.

The elongated forms of Africa and South America in particular make the pictures from Nimbus 5 seem so strange. Surprising as this may be, a look at a globe is sufficient to show that the unusual appearance is only the result of the projection, and has nothing to do with the actual shape of the continents. On a globe the continents appear as they actually are.

At the same time that the first pictures from Nimbus 5 were published, the German historian Arno Peters was struggling with the portrayal of the earth. After looking at a world map done in the Mercator projection, he had decided that the Old World, Europe, seemed oversized in comparison with the large areas of the Third World nations. This did not correspond to reality. He therefore promptly concluded that looking at such a map could cause Europeans to overemphasize their position in the world. Only one thing could help: new maps had to be made.

This is where Peters jumped in. He created a new projection, the Peters Projection, which gives us a more faithful portrayal of the actual areas of countries than does the 400 year old Mercator Projection of the German geographer Kremer.

What did this world map look like? Africa and South America had a strangely elongated shape similar to the way they appear in Nimbus 5's pictures. Peters believed he had finally found the "Egg of Columbus". He claimed that only his map gave a realistic picture of the earth. Peters even went so far as to ask the head of the West German radio network,

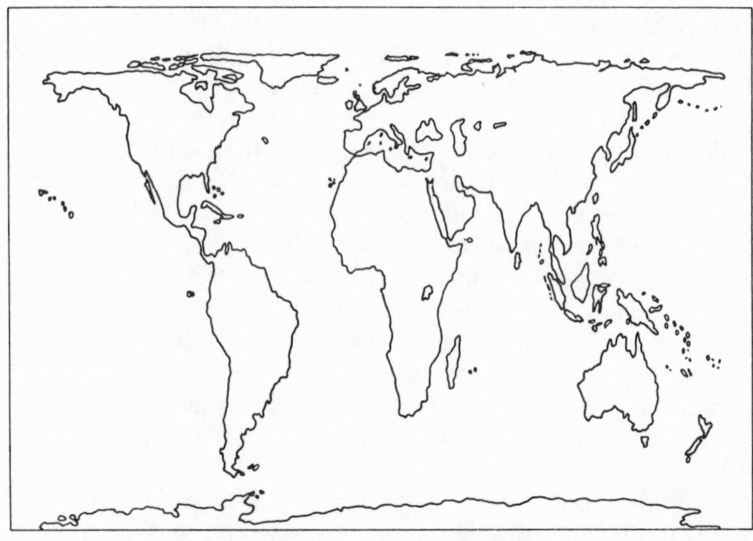

Map of the world by the German historian, Arno Peters. The regions near the equator show extreme distortion, most evident in the elongation of Africa and South America.

von Bismarck, to replace the world map with a Peters Projection on future television news programs.

The coincidence that Peters and Nimbus 5 "lived" at the same time, made this projection interesting for a while. In the long run, however, it cannot stand up. For one thing, the Mercator projection, which Peters wants to see banned, plays such an important role, especially in navigation, that eliminating it is inconceivable. For another thing, Peters is not the first to make an attempt to portray the surface of the earth accurately. Johann Heinrich Lambert also tried to do just that in the year 1772, 200 years ago. Since then, many other cartographers have imitated him.

Peters' attempts were doomed to fail at least partially, because there was only a very small need for a new projection. After all, it is well known that every map of the earth has inherent weaknesses. People are interested in the usefulness of a map, and do not pay much attention to the optical impression it makes. In spite of this, the world will remain whole: Africa is still Africa, Asia still Asia, whether on a Mercator, a Lambert, or a Peters Projection. Satellites emphasize this daily. The atlases based on the work of geodetic surveyors throughout the world are accurate. This realization should calm things down.

The Earth's Gravitational Field

Measuring the earth, creating a coordinate system on which countries and oceans are mapped, is surely an extremely important and many-sided task of the geodetic surveyors. However, it only remains on the "surface", so to speak. Everytime the press reports a terrible volcanic eruption or a devastating earthquake (like that which totally destroyed the capital of Nicaragua on Christmas Day, 1972) it becomes obvious that the earth possesses not only a surface, but also an interior where all kinds of things go on. Thus there is another task in measuring the earth: surveying the earth's interior. The irregular distribution of mass in the earth is responsible for many things that happen on the surface. It is exciting to realize how much information the orbits of satellites, high above the earth, can give us, not only about the invisible parts of the earth, but also about the movement of the continents, about volcanoes and earthquakes. The time is foreseeable when precise warnings of catastrophes will be possible. The final key to the secret has not yet been found, but that will soon change. Knowledge about our planet grows daily, and as soon as we have reached a certain point, we will make the big breakthrough.

Soon after the launching of Sputnik 1, the American astronomer Whipple organized a worldwide observation program in order to study the earth through its effects on the orbits of artificial satellites. Presumably he did not suspect what a powerful mechanism he had set in motion. Maybe these observations made him start thinking. The orbit of the first American satellite, Vanguard 1, proved to be extremely irregular. It was disturbed again and again by some unknown agent. More precise calculations showed that the resistance of the thin atmosphere still present at the altitude of the satellite was not responsible for this. Rather, an important factor was the noticeable flattening of the earth. The satellites made it possible to measure this with the greatest accuracy. Naturally scientists already knew of the earth's flattening. What was surprising was the realization that the earth is formed asymetrically. At the North Pole it is about fifty feet too large, and at the South Pole it is about fifty feet too small. This variation from symmetry is not great, but nonetheless it does exist. Apparently the earth looks like a pear, if we may exaggerate.

However, the real surprise was yet to come. On November 15, 1961 the Americans launched a navigation satellite, Transit 4B, which did not seem to follow the known laws of the universe at all. Sometimes, for no apparent reason, it sped up, and at other times it slowed down. Finally one scientist noted that it always increased its speed at the same place in its orbit, and it always slowed down at the same place. From this he concluded that not even the equator was circular; it too had bumps and indentations. Over the course of the years they discovered further bumps and indentations in the earth's gravitational field. For example, near Australia there is a low lying positive inhomogeneity of a material character, as the scientists call it. This means that mineral deposits under the earth's surface here are heavier than in other places.

This discovery is not only of theoretical importance. If these techniques of observation can be perfected, it will lead directly to "treasure hunts". Oil and mineral deposits have often been discovered near these indentations and bumps in the gravitational field which have been found in surveying the earth. Why this is so has long remained a mystery. The only thing we are fairly sure of is that this is no coincidence.

The Canadians have applied this knowledge, and have tested many areas from airplanes for irregularities in the earth's gravitational field. At the same time they have made magnetic and electromagnetic measurements, as well as photographs of the areas. In this way they discovered not only nickel deposits in Manitoba, but also other metals near Timmons, Ontario, for example.

Mapping the earth's gravitational field is thus a project that can be beneficial to all of us. However, the military are once again upsetting the plans. They are keeping their own results secret. Their measurements have to be much more exact than those of the civilian scientists, because it is their very substance that is involved. In calculating the trajectory of an intercontinental ballistic missile, much more is needed than just knowing the geographic coordinates of the launching point and the target. A trajectory is influenced by those variations in the earth's gravitational field which cause the rockets to speed up and to slow down, just as with the satellites. For this reason, the detailed structure of the earth must remain State Secret No. 1. There is no easy way out of this situation.

Fortunately, not all measurement of gravitation by civilians is forbidden. Thus, in spite of military secrecy, civilian scientists were able to expand our knowledge more and more, although at a somewhat slower pace. The most astonishing result they came up with was that the depressions and bumps in the earth's gravitational field coincided closely with the irregularities of temperature in the earth's interior. These irregularities are measurable with precise thermometers. Bumps, indicating heavy metal deposits, occur in conjunction with temperatures up to 50% lower than that of the surrounding area. Conversely, indentations, indicating light metal deposits, are connected with higher temperatures.

It was easy to conclude from this observation that the earth is extremely active in these places. Warm, and thus lighter metal deposits rise in the earth, while cold, heavy metal deposits sink down into the earth. And so we must further conclude that all mineral deposits in the earth's crust are in perpetual motion, because of differences in temperature. Alfred Wegener was probably right when in 1915 he maintained, for other reasons, that the continents of the world were nothing but swimming islands, constantly in motion. The final proof for this claim is still outstanding, but thanks to satellites, we now are coming closer to the truth. By using them, we can get an overall picture of the earth's gravitational field for the first time. The American satellite Geos C and the French satellite Starlette are primarily intended to provide the answer.

This whole area of research becomes interesting when we consider that the movements in the interior of the earth, the migration of masses of metals up and down, lead to "zones of weakness" on the earth's surface. These are associated with volcanoes and earthquakes, with the folding and shifting of mineral deposits – magnificent dramas of nature. Unfortunately, these dramas occasionally result in enormous catastrophes. In November 1963, the population was spared when Surtsey, an island born of fire and flame, arose in the North Atlantic, just south of Iceland. The floor of the Atlantic shifted, and Europe moved farther away from America. Nobody was injured, even though we can still clearly see the results of this upheaval in Iceland today. Broken-up chunks of earth are scattered over the entire island.

We cannot always count on having such good luck. The inhabitants of Central America in particular can tell another story. There, at the edge of the continent, the people are continually on the alert. The California peninsula is moving away from the continent, and, in the process, it is claiming its victims. There is never a restful moment along the weak zone, which extends northward to San Francisco, including all California's "faults", and southward to Nicaragua's volcano and earthquake region.

In 1906 San Franciscans woke up with a start one morning, as the earth began to tremble. Houses caved in, bridges collapsed, and hundreds of thousands of people lost their lives. Part of the earth had shifted twenty three feet in a single movement! Any day a similar misfortune could happen again, especially since, over the decades, much has been built right on top of these weak zones, ignoring the possible dangerous consequences. The Berkeley Football Stadium is built immediately over one of these faults, the Hayward fault – a direct challenge to nature. Only recently has construction sharply decreased.

Such shiftings of the earth are always going on. For example, the men who were working in an oil field in California were very surprised one day. All of their buried oil lines were twisted, some of them even ripped out of the ground. Was this sabotage? No, the earth had shifted. The subsurface was moving at a rate of about one and one-half inches per year.

Satellites are now surveying the critical areas of the earth. They are recording fault systems, which were often so concealed that they could not even be seen from airplanes. Movements in the earth's crust are registered. These, taken together with the measurements of the gravitational field made by the geodetic surveys and by satellites, give hope that one day all these processes will be understood. Then, finally, we will be able to warn against earthquakes and catastrophes. Meanwhile, more or less by chance, ways to regulate, or even hinder earthquakes have been discovered. A few years ago near Denver, Colorado an unusual rumbling in the earth's surface was registered. It was strongest in the U.S. Rocky Mountain Arsenal, where, at that time, poisonous liquids were being pumped deep down into the earth for disposal. As soon as the pumps stopped, the rumbling ceased – a strange phenomenon. Later experiments, based specifically on this observation, confirmed the phenomenon: weak earthquakes can be controlled with liquids. Will the satellites be able to fulfill their task of warning us? Will we live more securely in the year

2000, without fear of earthquakes? Only the future can answer these questions.

The Transit Program

When the satellite Sputnik 1 made its surprise appearance in the sky at the end of 1957, scientists all over the world were suddenly faced with a new problem. How could they continually track the orbits of the numerous satellites which would undoubtedly be launched in the next few years? Would optical observation be sufficient, or were other means necessary? The opinion at Johns Hopkins University in Baltimore was that optical observation alone would not be adequate. The physicists George Weiffenbach and William Guier therefore advocated a radio locating system, which seemed much more suited to the task. It had already been tested on satellites shortly before this by Wolfgang Priester, Hans-Gerhard Bennewitz and their co-workers in Bonn, Germany.

The two physicists started with the fact that the frequency of a signal increases as the transmitter approaches the receiver, and decreases as it moves further away. This is a well-known principle which can always be observed and which does not lose its validity in space. By using an appropriate receiver for the radio signals from the satellite, it would have to be possible to determine from frequency changes when the satellite was closest to the observer. This was indeed the case. They could even tell from the frequency curve how far away the satellite was from the observer at any given moment, and these data taken together permitted an exact determination of the satellite's position.

This insight could have satisfied the two physicists but fate decreed otherwise. A third physicist, Frank McGuire, joined the group. He had a simple, but ingenious idea: why not invert the process? If they knew the position of a satellite at every point in time, then they ought to be able to determine the coordinates of the observation station from the relevant radio waves. Based on this, ships, and, in certain cases, airplanes, when outfitted with the proper radio equipment and a computer to work the data out quickly, would be able to determine their position exactly. This was better than the processes used by the geodetic surveyors, which could not be applied very well to moving stations.

The idea was so good that the American Navy immediately entered the picture. A modern navigation system for their ships was precisely what they were looking for. Classical navigation methods were far too inaccurate. The Navy had long since abandoned astronomical navigation in favor of inertial navigation. In this system they take advantage of the fact that a fly wheel, once set in motion, always tries to rotate in the same plane. Any deflection is clearly measurable, so that course changes of ships equipped with such a system of fly wheels can be registered instantaneously. Taken together with the ship's speed, gathered from an independent source, the course and the position of the ship at any given moment can be calculated. They had to accept the fact that the determination of their position could be off by up to a mile and a half; otherwise how were the atomic submarines to navigate under water for months?

The new system of navigation by satellites resulted in an accuracy of ±100 yards. An extraordinary achievement! Now submarines could fire their rockets so precisely that they could reach any target anytime, because this new system was completely independent of the weather. Within two years, the Navy was far enough along to orbit its first test satellite. Transit 1B was launched on April 15, 1960.

Transit 1B had a predecessor. This was a hollow artificial moon measuring 195 feet in diameter and made of brick. It was the central point in a science fiction story by the witty author Edward Everett Hale, which appeared in 1869 and 1870 in the *Atlantic Monthly.* This story is especially interesting because here for the first time a satellite is mentioned in literature. Furthermore, this satellite was supposed to be used for navigation. Hale's idea was that the brick moon would fly around the earth at a low altitude. Its position could be determined at any time from tables, and, based on these data, one's own latitude could be calculated. This, together with the longitude obtained from observing the North Star, gave the ship's position at night. Naturally this brick moon never became reality. In practice it would have had many problems, but this was also true of the Transit program.

Three main problems arose. The frequency of the transmitter on board the Transit satellite had to be very well stabilized, because even small variations in frequency would lead to large errors. Secondly, the satellite had to have an exact signal to calibrate the time. Thirdly, and most important, the satellite's orbit had to be known exactly. This was the main problem.

Unfortunately, it does not help much to work with tables that give the exact positions of the satellites. Because of the earth's irregular gravitational field, the pressure of solar radiation and especially the resistance of the thin atmosphere, the orbits change so quickly that advance predictions of location gradually lose all meaning. After only sixteen hours the margin of error introduced in determining one's position is more than 100 yards. There is only one solution here, and that is that the ships must receive updated information constantly.

The system which finally crystallized looks this way. A series of observation stations scattered over the entire world continually track the Transit satellites. Every twelve hours their data are collated, and the most important orbital parameters are radioed to the satellite itself, together with the expected parameters for the change in orbit over the next few hours. These data are then rebroadcast by the satellite so that every ship that wants to determine its position can compute where the satellite and the ship itself are at any given time.

Because of the complicated calculations required to find the true position of the satellite, the amount of apparatus required on board the ship is naturally considerable. Since a

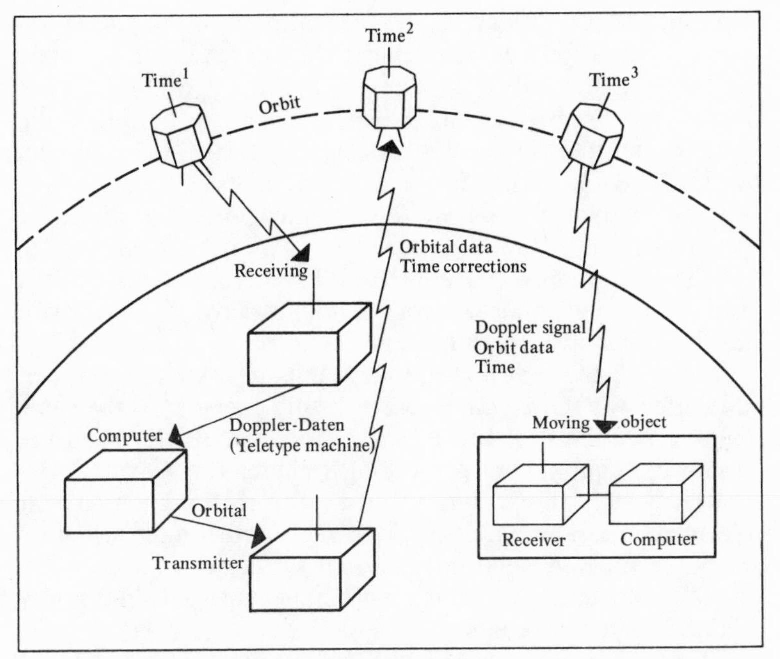

A schematic diagram of the Transit navigation system. First the orbit of the satellite is precisely calculated. A ground station then transmits the necessary orbital data to the satellite itself which relays them back to earth as it continues on its orbit. Taking this information together with its own radio observations any ship can determine its position.

satellite navigation system costs several thousand dollars, not every ship is equipped with one. This was not the initial intention anyway. The American Navy was not even thinking of making this system available to civilian ships. Besides that, they had to struggle with technical difficulties for a long time. Again and again there were failures and setbacks.

In November, 1960 the Transit program unexpectedly claimed a victim which, because of the unusual circumstances involved, was ceremoniously buried. The entire village of

Hulgin, Cuba took part in the funeral procession, and everyone's sympathy could be felt in the speeches. A few days earlier, on November 30, the second stage of the booster rocket for Transit 3A had failed, and had to be destroyed. The ruins had fallen fatally to the earth near Hulgin, killing a pig – a victim of navigation satellites.

At the end of 1964 the satellite navigation system of the American Navy, the U.S. Navy Navigation Satellite System (NNSS) was finally ready for use. Three years later, in July 1967, President Johnson gave the long-awaited signal to open the Transit system to civilians.

One of the first non-military ships to take advantage of this offer was the German research ship Meteor. At the same time it was the first European ship to be outfitted with the necessary equipment. After long debate, the German Research Society (DFG) had voted to absorb the costs so that German ocean research could carry on the most accurate work possible. A good navigational system is extremely important in charting the ocean bottom, because only in this way will the results be reliable.

The Need for New Navigation Systems

Commercial shipping as well as air transport have fought against using the Transit system. The main reason for this is probably that the receiving and computer equipment necessary on board the ship or plane are too expensive. Undoubtedly there are other reasons. The need is certainly great enough, although future business travel is supposed to decline, in spite of better transportation facilities. This is astonishing, but not too long ago the trend was verified in an experiment.

The Westinghouse Corporation carried out the experiment. The company, which has numerous widely scattered branches, determined that trips by management personnel demanded a larger and larger budget every year. The director's convention alone was an expensive undertaking.

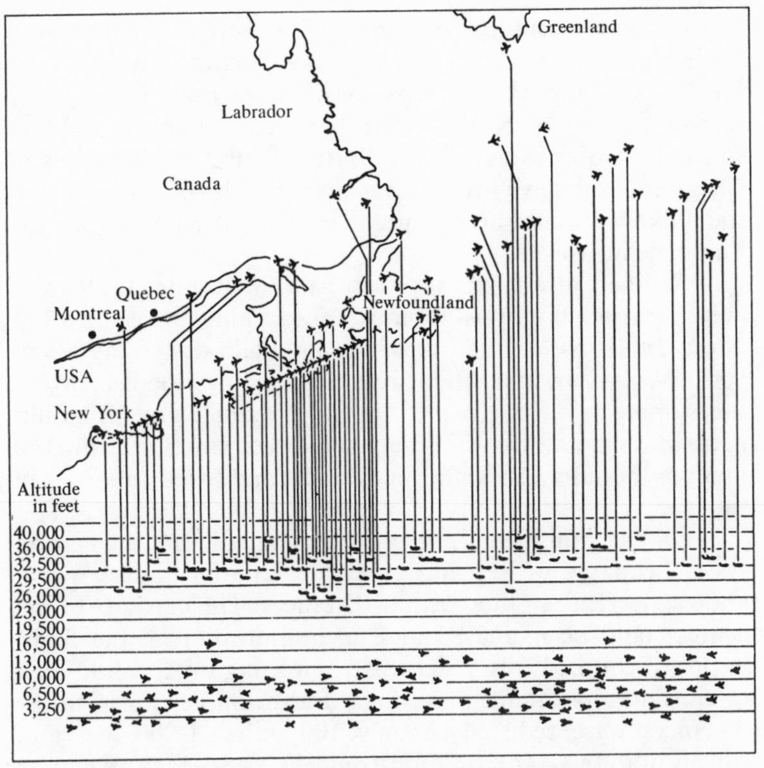

An example of the density of air traffic today: the chart shows traffic across the North Atlantic on June 18, 1966 at 9:30 P.M. EST. The conditions have become even worse in recent years.

For that reason they made the best means of communication in the world, including satellites, available to management. Lo and behold, business trips decreased by an astonishing 20%. They had become superfluous. Many people consider the test a contribution to the battle against pollution – telephones are nonpolluting!

This experiment does not alter the fact that tourism and transportation of goods are continually on the increase, a sign of the rising standard of living in the world. According to fairly recent calculations, the experts expect that there will be 18–20,000 ships with a capacity of more than 1,000 gross register tons by 1975. At present there are probably over 3,000 ships of this size sailing on the North Atlantic simultaneously. Cheap satellite navigational systems would be just as useful to them as they would be to a number of smaller vessels, for example fishing fleets.

Airline traffic over the North Atlantic is indeed less than ocean traffic, but it is rapidly approaching the saturation point. In 1976 the number of flights made daily may reach 900. Even today some flight routes are so overloaded during the summer tourist season that only a small growth would necessitate increasing the safety distances, which are now 140 miles to the sides, 2,500 feet in altitude, and twenty minutes in time.

On the other hand, an appropriate satellite navigation system capable of keeping every airplane on course would allow a decrease in these safety distances. The capacity of the routes could be increased and an optimization of the flight patterns, coupled with a decrease in waiting time, would lead to considerable financial savings. For example, if the safety distances were reduced to only 100 miles, 1,000 feet and fifteen minutes, $46.5 million could be saved each year over the North Atlantic alone.

According to NASA's calculations the financial advantages of a satellite navigation system would be even greater in shipping. On the one hand, ships could maintain their courses even more closely, so that they would use less fuel; on the other hand, they would have to spend even less time on astronomical navigation, which consumes approximately 380 hours per ship every year! By saving only 1% in these two areas, shipping would gain $150 million per year.

Naturally these figures were sufficient to motivate a search for other satellite navigational systems using cheaper equipment on board. The American Navy took the first steps towards this on September 2, 1972. On that day it launched a new type of navigations satellite bearing the name Triad.

The nucleus is a small one inch sphere covered with gold and platinum, developed in the Laboratory for Applied Physics at Stanford University. This sphere floats freely in a vacuum container. Every time the orbit of the satellite is changed even slightly by the influence of the atmosphere, there are disturbances. These in turn influence the position of the vacuum container, but not the position of the ball itself, since it is completely cut off from the outside world. Because of this the sphere hits its container and the collisions cause rockets to be fired to correct the course. This then eliminates the influence of the surrounding atmosphere and the pressure of solar radiation. The satellite Triad moves in an orbit which is almost completely free of outside influences.

This is precisely the advantage of the Discos system (Disturbance Compensation System). An undisturbed orbit is much easier to calculate than an orbit with many perturbations. Vessels that want to determine their position using Triad can get by with much cheaper navigational equipment than they could using the Transit system. This was the first achievement.

Meanwhile the American Air Force was developing a navigational system conceived along completely different lines; it was based on stationary satellites. This navigational system only works if the Doppler measurements of the Transit system (which give usable results only at a fairly high relative speed) can be replaced with measurements of distance, for example, by the already mentioned Secor process. Other methods are also possible, whereby only the vehicle itself sends out radio signals which then reach a ground station via the satellite. Here they are evaluated and the position of the vehicle is radioed back, again, via satellite.

With the help of an ATS satellite, tests were carried out in 1968 demonstrating the usefulness of such a geostationary system. A completely ordinary automobile drove from Baltimore to Washington. It differed from other cars only in its "payload". It carried a 25 pound package, a radio transmitter which continuously broadcast radio signals. These signals reached Greenbelt, Maryland via the ATS satellite. At the Goddard Space Flight Center, NASA could keep a running check on the position and speed of the automobile. Similar experiments were performed with ships and airplanes. This so-called "external locating process" has the advantage that the satellites no longer only determine position, but they can also relay information at the same time. In this way ships and airplanes receive the latest weather and environmental data. If there should be a shipwreck, rescue attempts could be started immediately, because the position of each and every vessel is known at a central location.

Indeed the principle of supervising all shipping is coming more and more to the fore. Aeronautical satellites, originally intended for air traffic, are assuming a greater and greater role in future planning. We are approaching traffic supervision from outer space. The Americans and Europeans have recognized the signs of the times. In 1971 in a "memorandum of understanding" they agreed to develop jointly such a satellite. Yet as so often, pettiness hindered progress. In February, 1972 President Nixon refused to sign the agreement. American industry, not wanting to let Europe take part in space exploration, had gotten its way.

The Europeans were shocked. In March, 1972 they expressed this shock to the American negotiators who came to London, Paris, and Bonn trying to make bilaterial agreements with the individual countries concerning an aeronautical satellite. Europe remained obstinate. Consequently, the delegation from the United States returned home with its task uncompleted – without even going to the capitals of the other European countries as originally planned.

In the meantime the International Maritime Commission, IMCO, has shown interest in traffic supervision from outer space. In 1972 it suggested a common system for air and sea travel. This irritated the Americans because the Soviets now took up the issue. They had long kept silent about the U.S. influence in the communications satellite organization, INTELSAT. There the Americans do have the most influence because their needs for communications are greatest.

In shipping things are different. Here the Soviets set the tone, and the Europeans are following their lead. The countries belonging to ESRO (European Space Research Organization), together with Norway, make up about 40% of the world's shipping tonnage, while the United States have been forced into a minority position, with only 10%. Are the Soviets getting their revenge here for their defeat in satellite communications? Or has the suggestion by the American COMSAT to cover maritime travel with its own communications satellite system acted as a buffer? The battle of the giants has not yet been decided. One can only hope that general interests are not completely lost from sight. The time is ripe to make air and ship traffic safer through satellites.

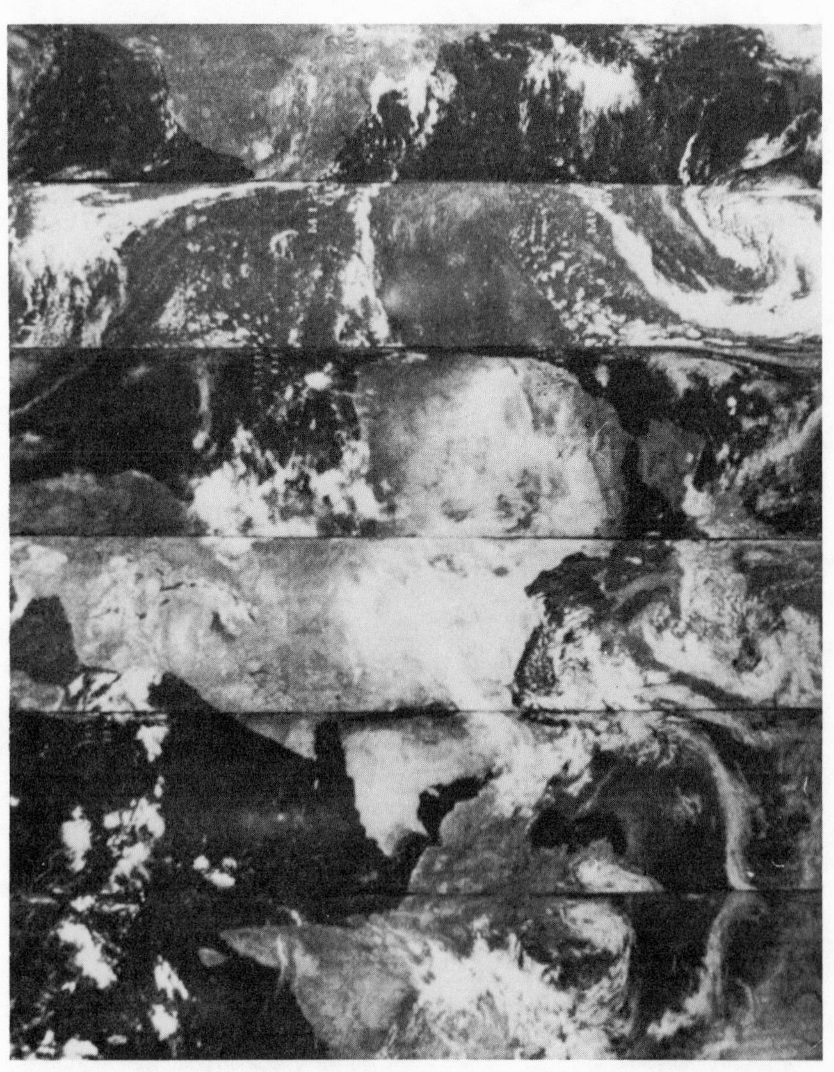

Thermal infra-red pictures are very similar to photographic pictures. White cloud fields are sharply differentiated from the darker background. A look at Africa from Nimbus 3.

6. New Directions in Weather Research

A Few Words About the Weather

It was mid-winter when the weatherman's telephone rang. Not suspecting what was in store for him, he answered his phone, and a flood of words from an elderly lady filled his ears. Apparently she did not agree with the weather forecast for the previous day. Annoyed, she complained, "I wish you would come here and shovel these six inches of 'light clouds' out of my driveway". In spite of the most painstaking evaluation of all weather observations, this meteorologist had made a mistake in his forecast.

There is scarcely a topic more talked about than the weather. After all, we are always in contact with it, and many things depend on it, not the least of which is our mood. Why should it surprise us that questions about tomorrow's weather are asked so often? An army of meteorologists throughout the world tries to answer these questions daily. Nonetheless, they have not succeeded, even to the present, in producing predictions that are 100% reliable. Because of the number of factors which have to be taken into account, precise calculations are impossible.

In principle, it is all quite simple. Let us take a step-by-step look at the most important factors influencing the weather. Let us assume that the earth consists of the same surface material at all places, that its topography is the same everywhere, and that it does not rotate on its axis. What happens then? A global circulation of the atmosphere in a north-south direction occurs.

The reason for this is that the sun's radiation does not strike our home planet at the same angle everywhere. The angle is steeper in regions near the equator than at the poles. This means that the land is warmed up more at the equator, and, also, that more warmth is radiated back. As a result of this radiation, the atmosphere, the gas casing that surrounds the earth, heats up, and consequently rises. A "hole" develops in the lower levels, especially near the equator. This then has to be filled somehow. Cooler air streams down from the polar regions, and this too has to be replaced. A counter-current of warmer air at greater altitudes arises, and a perfect circulation is achieved.

This simple consideration alone is enough to show how important the sun is for the earth's weather. It is our number one source of energy. Not only is it the precondition for all human and animal life, for the growth of the plant world, and for many deposits in the earth (including oil and coal as the remains of long dead flora), but without it, the weather would be incomprehensible.

Unfortunately, the conditions are not as primitive as this. In reality the earth rotates on its axis. This is the second factor which we have to take into account. It causes the so-called Coriolis force: a point at the equator rotates faster than a point at a higher latitude, because it has to cover a much greater distance in the same amount of time, e.g. 24 hours. The same thing is true of atmospheric particles.

When air from the equator flows toward the north or the south, it retains its original speed of rotation, and thus moves too quickly. This means that it is diverted from its north-south direction, to such a degree that in our simple model the air currents can only reach to about 35° latitude. That is where the equatorial wind-system ends.

Naturally the cool polar winds continue to blow, in the south towards the north, and in the north towards the south. Similar considerations are valid for them too. At some point, the polar wind-systems collide with the equatorial wind-systems. Already things are becoming more complicated, although we have only considered two factors. We do not even need to point out that most of the United States also lies in this "collision zone" in the medial latitudes between 35° and 60°.

Up to this point we have not even made mention of other phenomena such as rain, snow, hail, and clouds. These are traceable to the chemical composition of the earth's atmosphere, and consist partially of water vapor. It is well known that warmer air can hold more water than colder air. When this warm air flows into a cooler region, its water vapor content becomes too great—the experts say that it is supersaturated. Consequently, the excess water vapor condenses, and forms either fair weather clouds or rain clouds. The meteorologist is now faced with a problem: how can he predict when clouds will develop, or when they will become so large that they burst, covering the land with rain, snow, or hail?

Things would certainly be much easier if the earth's topography were the same everywhere, as we initially assumed. Fortunately, this is not the case; if it were our earth would be boring. Who does not like to dive from the shore into the cold waves of the sea? Who does not like to look down from a mountain peak to a green valley? It is exactly this richness of variation which makes our weather so interesting.

The differences between land and sea contribute to multiple atmospheric whirlpools. During the day the land is warmer than the sea, so that the air over it is warmer. This warmer air then starts to circulate, to the grief of the meteorologists. At night it is reversed. The warmth taken up by the sea during the day is radiated back in to the air more slowly than that absorbed by the land. A nice sea breeze often makes us aware of this. The same effect is even felt near fairly small lakes.

Mountains are another factor affecting the weather. Stretching up into the sky, they stand in the wind's way, creating windless areas and fall winds of the most unpleasant kind.

Every pilot can tell stories about them. The situation which was so clear at the beginning is basically now so confused that we are still not able to calculate the next day's weather accurately from pre-existing weather conditions. The meteorologists are slowly approaching this goal, but they will only be able to achieve 100% success when they have sufficient observations of wind direction, wind speed, temperature, air pressure, moisture, and other weather phenomena throughout the world. Hopefully they will then be able to fathom the laws of weather.

Presently meteorologists have to be satisfied with predicting the development of the weather by comparing the current situation with earlier ones. Their prognoses are based on experience. Here too, it is essential that observations from a sufficiently large area be included, since, as we have seen, the weather at one place on the earth is influenced by the weather in distant areas. The advances made in weather prediction have come about to a large extent because of space exploration. It has made the process more exact, and thus has become irreplacable for those branches of the economy dependent on weather.

An Idea Matures

The realization that satellites could provide a valuable service in weather prediction is quite young. As recently as 1954 almost nobody believed in it. Harry Wexler, the head of the U.S. Weather Bureau, considered the idea so crazy, that he agreed to Arthur C. Clarke's suggestion that he prove how ridiculous it was. This brought about the turning point; he failed to prove anything—to the contrary, Wexler suddenly became enthusiastic about the idea. In fact, he became the foremost proponent of giving satellites a role in predicting the weather.

The meteorologist's original aversion to weather satellites orbiting the earth is not as incomprehensible as it may seem to us today. We must bear in mind that weather actually takes place almost solely in the lower levels of the atmosphere, in the troposphere, which varies from five miles deep at the polar regions to ten miles at the equatorial regions. Satellites can not fly at this altitude. If they are to stay aloft for a fairly long time, then they have to be several hundred miles above the earth's surface, that is, far beyond the actual region where our weather develops.

It was not immediately obvious that weather observations from these heights could be useful. To be sure, the meteoroligsts knew even at that time that clouds could be clearly identified. However, they did not know clouds are so dominant, that they cover entire countries and continents. Many scientists did not believe either that they could see enough details in the cloud pictures from satellites to reach conclusions concerning the weather as it was developing. The way things turned out, useful evaluation of these pictures was only possible after a new classification of the different cloud types was introduced. The global overview of the earth provided by satellites turned several areas of meteorology upside-down.

In 1954 there was no way for satellites to measure various weather phenomena such as temperature, air pressure, wind speed, and so forth. They were definitely at a disadvantage compared with weather balloons and high altitude research rockets. Although this development too was at its very beginning, these devices had already been used occasionally to investigate the secrets of our atmosphere.

The idea of using rockets for weather research had been under discussion for years. German scientists had suggested the V2 rocket, actually used as a destructive weapon, be used for weather research. As is well known, the course of events took a different direction. Germany lost World War II, and all rockets were shipped out of the country. Scientists in the United States took up the idea only hesitantly. Test shots were made, and the new knowledge they provided was so phenomenal that no one wanted to do without these rockets any longer.

Nonetheless, investigating the weather with the help of high altitude rockets progressed slowly. In the U.S. a regularly employed meteorological rocket network developed from a suggestion by Hans aufm Kampe in 1959, but these rockets only measured wind speed. Other measurements were added later in the sixties. Satellites at a high altitude were unsuited for determining wind velocity. So what were they supposed to do?

It is a mark of Harry Wexler's deep insight that he so strongly espoused perfecting weather satellites in spite of objections. At the very beginning of 1959 the first primitive test was carried out. On February 17 the satellite Vanguard 2 reached orbit around the earth. In addition to several other scientific instruments, Vanguard 2 carried two telescopes, which were supposed to photograph the earth's cloud cover. The principle behind this photographic technique is so ingenious that we will explain it briefly. In a similar, although perfected form, the technique is still used today for pictures of the earth in the infrared and radio ranges.

The main component of the photographic equipment was two normal photoelectric cells. These change any light striking them into electrical current. The photoelectric cells were fastened to the outside of the satellite. In front of each of them was one of the telescopes, which meant that light could only strike them at a very sharp angle. This was all there was to it. The technicians just had to orbit the satellite, and then make sure that its axis of rotation pointed in the direction the satellite was moving. Now, when the satellite rotated on its axis, the telescope scanned a narrow strip of the earth, so that the photoelectric cells could register the precise brightness value of this strip.

The satellite meanwhile continued to rotate. During its next rotation, it scanned another strip of the earth which was not identical with the first one. The satellite, because of its own orbit around the earth, had moved forward a bit. It the speed of rotation were correlated with the satellite's own movement,

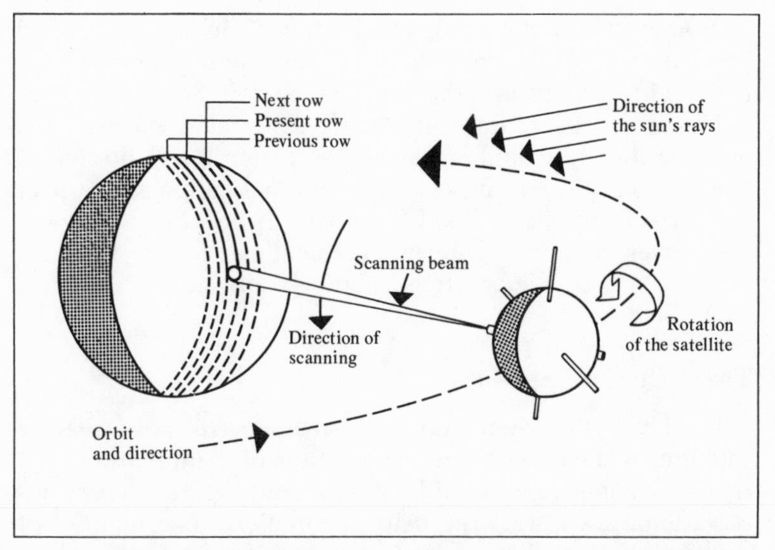

A schematic diagram of the photographic methods used to obtain the first pictures of the earth via satellite.

then the strips of the earth, whose brightness relationships were being recorded, would be adjacent. From the electrical signal which the satellite sent back, a picture of the earth could be produced, line by line, just as in television.

This was the theory. In reality, things went differently. The Vanguard satellite was true to its series. (In spite of the name Vanguard 2, it was the seventh of its family.) The technicians were happy it had reached its orbit, and did not have to be destroyed like most of its predecessors. On the other hand, it led its own willful life, and disappointed all hopes for pictures. Its orbit was by no means circular, its speed of rotation was much too small, and, besides that, it executed such uncontrolled motions that the project failed. No pictures of the earth were made.

The scientists did not give up. The Army Signal Research and Development Laboratories, which had designed the equipment, did not diminish their efforts. On August 7, 1959 they launched Explorer 6 with the same photographic equipment. This satellite accomplished its mission. Seven days later at about 11,100 miles altitude, its equipment received the order to start functioning; shortly thereafter, the radio signals reached Hawaii. For the first time they had gotten a picture of the earth from a satellite. This picture is now history. It was taken over Mexico and shows a crescent earth scarcely recognizable to the layman. It is shadowy, and covered by vast cloud fields.

The Great Stroke of Luck

In 1959 the meteoroligists' attitude toward weather satellites had changed somewhat. Many of them had realized that they ought to make at least an attempt to gather new observational data with the help of satellites. The number of conventional weather stations was much too small to get an adequate overview of the weather. The network had too many large gaps. This is true even today. While Germany, for example, has one such station for every 3,900 square miles, several of the developing nations have only one for every 20,000 square miles.

The situation on the oceans is catastrophic. There are automatic weather buoys scattered here and there, and commercial ships are in contact with the weather observation network. These are only a drop in the bucket, however, especially since many storms gather over the oceans unnoticed, and then strike the coasts, where they then have disastrous results. Even weather balloons and high altitude rocket research are not too effective. In 1959 the observation network was much too wide-meshed for an up-to-date weather forecast, which presupposes a knowledge of current weather conditions in the entire world.

As is so often the case, it was the American Military who were first openly in favor of this new development. Even before Sputnik 1, various branches of the military had started constructing orbital weather reconnaissance systems to help them in international operations of all kinds. In mid 1958 the Department of Defense put ARPA (Advanced Research Projects Agency) in charge of all existing projects. This agency designed a satellite which was to be named Janus.

No one would have been surprised if the military had orbited their first weather satellite a year or two later, but they amazed everyone. In April 1959 they handed the project they had started over to NASA. Apparently they did not have too much faith in their own weather satellites, because in the following period there was unusually active cooperation between the military and the civilians. The military relied on NASA's satellite pictures for their operations, even as late as the mid sixties.

NASA quickly set about completing the project. Together with the Army Research and Development Laboratory for Communications Technology, they worked in high gear, and within one year they were ready. On April 1, 1960, they launched the weather satellite Tiros 1, which had developed out of the Janus project. This introduced a satellite series which even today is considered to be the most successful ever. The Tiros satellites opened up new dimensions. In the truest sense of the saying they helped to "move mountains".

Tiros is an abbreviation. The word stands for "Television and Infrared Observation Satellites." This makes clear immediately what tasks were set for the satellite. For one thing, the Tiros satellites were to test whether satellite pictures of the earth could be used for weather prediction, and if so, how. For another thing, they were to carry out those measurements in the infrared range, so necessary for making inferences concerning the energy content of the earth's atmosphere. Tiros 1, however, did not have any infrared equipment on board; the meteorologists were initially satisfied with observing the earth and its cloud distribution.

Naturally, they did not immediately expect to obtain crystal clear pictures on which every tree and every bush could be identified. Tiros 1, like its followers, flew much too high for this (about 450 miles). In addition, it was supposed to function for several weeks. Thus it did not have any normal photographic equipment, since the film would have taken up too much space. Instead, the satellite had two television cameras whose pictures were stored on magnetic tape for rebroadcasting to the earth at a later date. Direct broadcast was only possible over a few receiving stations, and was thus of minimal importance.

The process of television relaying was complicated. This meant that, during the initial stages in the use of weather satellites, it was not possible to regularly include cloud pictures in the daily weather forecasting service. Although there were exceptions, the first Tiros satellites were primarily used to test a new technology— a technology which very soon bore its first fruits.

One of the main drawbacks of the satellite was the fact that its magnetic tapes had a limited storage capacity. It was not possible to get more than 32 pictures of the earth per orbit. These pictures were transmitted to one of the CDA stations (Command and Data Acquisition). As soon as all the data were played back, the photographic equipment on the satellite was re-programmed, and given instructions when to resume photographing.

It was not possible to program the Tiros satellites so that they would, for example, take two pictures over North America, three over Africa, and the rest over Asia. All that could be controlled was the timing to set the cameras in motion. Once this moment arrived, the photographing and storing of the 32 pictures followed in a predetermined sequence. In a quarter of an hour the process was completed. For the remainder of the circuit around the earth the cameras were silent.

Unfortunately not all the pictures taken in these fifteen minutes were good. The Tiros satellites were torque-stabilized. That is, by rotating on their own axis, they maintained their orientation in space. The cameras were mounted parallel to this rotation, so that they too always had the same orientation. Because of this however, they were not always aimed directly at the earth. Consequently, some of the photographs so distorted the earth that evaluation of the pictures was not even considered.

A global recording of the earth was also made impossible by the fact that the satellite's orbit gradually shifted in space. Eventually the satellite flew over the Northern Hemisphere for weeks only at night, and then for the next few weeks over the Southern hemisphere. The Tiros satellites did not yet have night cameras. Furthermore, their orbit was inclined 48° (later 58°) to the equator, so that the picture only showed recognizable land and water masses to approximately 60° or 65° latitude. For all of these reasons, the satellites photographed just a portion of the earth's surface. Only later did truly global observation become feasible.

Thus the Tiros system served mainly for research and the testing of a new technology. In order to obtain the greatest possible advantage, the Americans had asked meteorologists throughout the world to let them know what parts of the earth they wanted photographed. Everybody interested was regularly informed which areas would be within range of the cameras over the next few days. The requests for pictures based on this information were compared and submitted to experts. Finally, the National Weather Satellite Center of the U.S. Weather Bureau, Suitland, Maryland, decided when the cameras were actually to start photographing. Special requests were also accomodated. The U.S. and Canada were pursuing an ice reconnaissance program supported by the Tiros satellites. Furthermore, scientific expeditions were given access to the pictures on demand, the International Indian Ocean Expedition and the Equatorial Atlantic Expedition, for example.

Given the limited range of possibilities, the planning was fantastic. There was only one point that the scientists had apparently estimated incorrectly: the volume of the resulting data. They still did not suspect anything when Tiros 1, at the end of its first revolution, made astonishing pictures of the St. Lawrence River in Canada. Yet the sobering up quickly followed. Every week 4,000 new photos came in, and after only a few weeks, when Tiros 1's cameras ceased functioning, the number of pictures had already grown to 23,000. There are now more than one and one-half million weather satellite pictures. A precise evaluation of these pictures is almost impossible, because of the volume.

On the Way to New Discoveries

The scientists were faced with the problem of trying to determine which part of the earth the satellite pictures actually portrayed. This was easy when characteristic borders and outlines were recognizable, but that was only rarely possible. Clouds usually obscured such details. Analysis would have been meaningless unless the problem of determining exact location could be solved.

Gradually, progress has been made. From the very beginning, the radio command to start the cameras could be used to decide what part of the earth was being photographed. This information established the time of photographing exactly, and, from this, the location of the satellite could be calculated without any trouble. In addition, there were methods of computing its altitude, as well as rotational position. Later Tiros satellites also carried equipment capable of determining true North. With the help of computers the problem of orientation was finally eliminated.

Computers also aided in "explaining" the pictures. After being fed the relevant data, they automatically drew a geographic grid which was then superimposed on each individual picture. At the same time the photographs were adjusted. This was required since television cameras cause unavoidable distortion in reproduction. The influence of the earth's curvature and the sloping position of the satellite in space at the time of photographing could also be adjusted, as long as these were not too great.

After this preliminary work, the actual analysis of the pictures could begin. This was done mainly in the Command and Data Acquisition stations at first. The most important of these are in Ft. Monmouth, N.J., Kaena, Hawaii, Wallops Island, Va., Fairbanks, Alaska, and Point Mugu, Calif. Only two or three of these stations were used per satellite, and some of them were only built later.

Analysis proved to be extremely complicated. Not much of what is visible from the earth can be recognized on the satellite pictures. For example, the world organization of meteorologists had made a classification of ten different types of clouds. Only three of these were rediscovered on the pictures. The rest of them did not show up. Either these clouds disappeared due to the resolution of the picture, or larger structures forced the details into the shadows.

The meteorologists were originally unable to make weather forecasts on the basis of satellite pictures. As a matter of fact, they did not really know what to do with them. Only by comparing them with conventional weather maps could they say anything intelligent at all about them, in spite of the fact that they wanted to use to satellite pictures in drawing up new weather maps. This unusual situation foretold that a new science was being born. Today, things are different. There are quite a few meteorologists who can tell from satellite pictures alone how the weather is developing, although these pictures do not include all the data found in weather maps.

Within just a few years new bits of information concerning our weather were gathered. A new area of research opened up for meteorologists. Suddenly they saw a completely different, unusual world that they had not even dreamt existed. One layer of clouds after another moved with the wind. Where there seemed to be a gap in this system, the television camera showed yet another layer of clouds. The most noticeable thing was probably the great number of spiral shaped structures, whose extent no one had suspected at all.

It was not especially surprising that cyclones exhibited such structures. What was surprising was the magnitude of the spirals. Near Brisbane, Australia a spiral with a diameter of about 1,100 miles was discovered on the satellite pictures. These spirals were not associated solely with cyclones. From the Bermudas, for example, a ship sent out a weather report, on the basis of which an extensive shower area, but nothing more, was drawn on the weather maps. One of the Tiros satellites made the meteorologists sit up and take note: it showed a previously unsuspected cloud spiral measuring 400 miles across.

The meteorologists were just as astonished when they discovered some enormous spirals on their satellite pictures at places where some time before, low pressure areas had influenced the weather. These areas had, in their opinion, long ceased to exist, and had disappeared from the weather maps. There are many examples of how incomplete much of our knowledge concerning weather actually used to be, even though we should not disparage the "pre-weather-satellite meteorologists". Many areas of meteorology had already been investigated so thoroughly that the satellites contributed nothing new.

By 1962 the scientists had become so accustomed to reading the pictures from the Tiros satellites, that they could establish a new catalog of cloud types. J.H. Canover initiated this work. He created the system in which clouds are classified according to brightness, formation, color, height, shape, and size. A year later the catalog was expanded once more, and the important principles for evaluating satellite pictures were finally established. A milestone in satellite meteorology had been attained.

Shortly after the launch of Tiros 1, weather satellite pictures proved how valuable they could be. For this reason, the meteorologists tried to find ways of placing the photographic material at the disposal of local weather bureaus. Transmitting the pictures themselves to all parts of the world was not even considered, since only very few stations possessed the necessary receiving equipment. They had to satisfy themselves with broadcasting coded data into the ether for interpretation by interested meteorologists, who then included it into weather maps. This naturally took a great deal of time.

American scientists then returned to an old method which had been almost completely forgotten: nephanalysis (cloud analysis). Nephanalysis is basically nothing more than a simple description of the clouds. It is included in weather maps in the form of signs and symbols. During the last century it was quite widespread, especially in Anglo-Saxon countries. The CDA stations now made such nephanalyses, and broadcast them to the National Weather Satellite Center in Suitland. Later the Center took over the task of making nephanalyses itself.

Nephanalyses proved to be so valuable for weather prediction that on April 15, 1962 they supplanted coded radio weather broadcasts. From this time on they were regularly transmitted with other weather broadcasts. The nephanalyses were sent from Suitland over the wires of the American Facsimile Picture Network to New York and San Francisco. In this way they reached all the American weather stations belonging to the network.

From New York and San Francisco the government-owned radio service broadcast them to the rest of the world. The radio stations in New York reached the weather centrals in London, Paris, Frankfurt, Rome, Cairo, Budapest, and Moscow. Those in San Francisco reached Mexico City and Cape City, Dakar, Senegal, and many other countries in Asia and Australia. Soon a hundred nations were part of this distribution network. Nonetheless, pictures from the Tiros satellites were rarely used in daily weather forecasting. As a rule it took six to eight hours from the time the photo was shot before the weathermen received it, much too long to be of any use. In addition, the Tiros satellites still could not photograph the entire earth.

Eventually, however, all these problems were worked out.

Using cameras whose pictures were continously relayed reduced the amount of time to the final consumer drastically, and enabled every country in the world to receive them. Meanwhile improved orbiting and orientation techniques facilitated a comprehensible coverage of the earth. Tiros 9, launched on January 22, 1965, initiated these new techniques. We should mention that only the large American CDA stations can receive the pictures from Tiros 9. We will talk about the direct broadcast of pictures in another place.

Tiros 9 is the first of a series of satellites called "wheel satellites". On the outside it is very similar to its predecessors. It owes its nickname to its orientation in space. Like Tiros 1-8, it too was torque-stabilized. However, Tiros 9's axis of rotation was not in the plane of orbit, but was perpendicular to it. Thus the satellite "rolled" around the earth. This had the following advantage: the cameras could now be mounted on its exterior, and every time the satellite rotated on its axis once, the cameras were pointed straight at the earth. Precisely at this moment the camera shutter opened. In this way, the technologists eliminated the great distortion in the pictures of the earth.

A further improvement in the photographs was reached by orbiting Tiros 9 at an inclination of 81° to the equator. The satellite now crossed the earth in an almost perfect north-south direction, and even the poles lay within the range of observation. Indeed the orbit was calculated even more precisely: the technicians caused the satellite to drift exactly one degree to the west every day. This negated the earth's movement around the sun—also a matter of one degree per day. While photographing was going on, the sun always stood at Tiros 9's back. The scientists thus achieved the most favorable conditions for observation.

There were still two improvements necessary to photograph the earth completely, within a relatively short amount of time. These were not included in Tiros 9. It did not have any night photography instrumentation for round-the-clock observa-

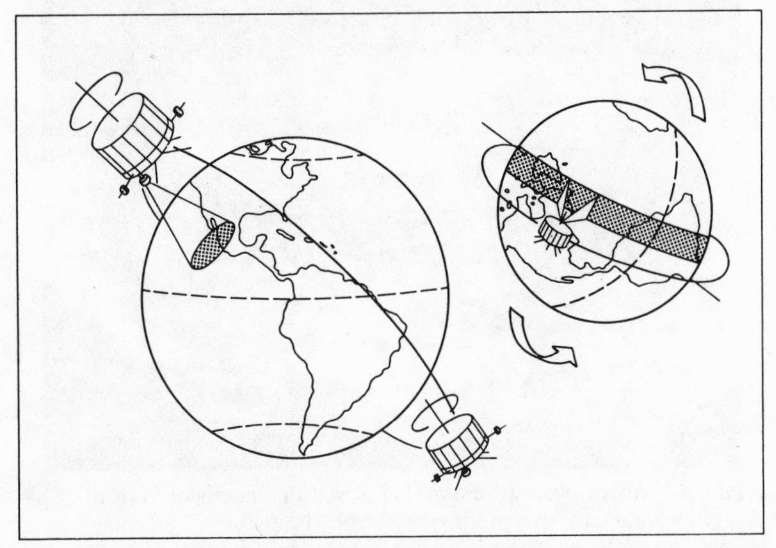

Because of the orientation of the axis of rotation and the mounting of the cameras, a large percentage of the photographs of the earth made by the first Tiros satellites were strongly distorted (left). The "wheel" satellites" of the Tiros series, on the other hand, took photographs that were almost totally without distortion (right).

tion, and its magnetic tape equipment was inadequate. Under favorable conditions enough photographs for a complete survey could not be stored for later retrieval by the CDA stations. (The orbit of the satellite was strongly elliptical.) Tiros 9's followers had enough storage equipment for a total of 48 pictures. They also revolved around the earth at a greater altitude, so that this problem too was solved. The global observation system is now a reality. Twice daily—once during the day and once during the night— every point on the earth is photographed, and the pictures are evaluated by meteorologists throughout the world.

Above: A storm over the Pacific, 1,250 miles north of Hawaii. The Apollo 9 crew took this picture in March, 1969.

Below: Early hurricane warnings and measures against these hurricanes are often possible only with the help of satellite pictures. This picture shows the hurricanes Beulah, Chloe, and Doria, which were seeded in 1967.

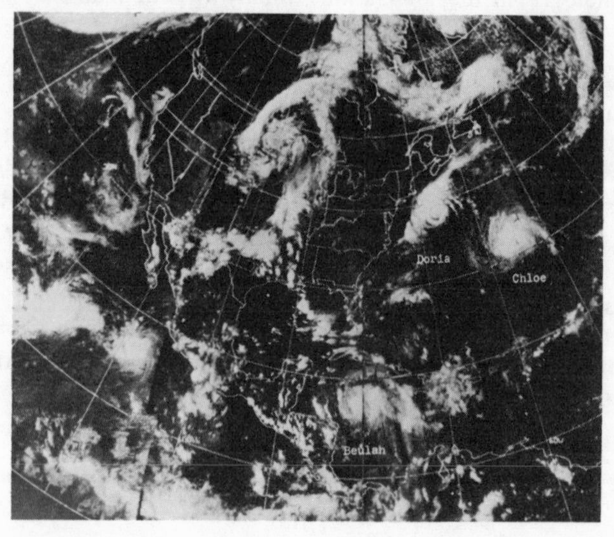

7. Satellites Save Lives

Camille

August 1969. The most recent pictures from one of the Essa satellites had just arrived. They were immediately placed in the hands of the meteorologists, who examined them from the expert's point of view. At first glance, they did not notice anything, but then they saw it: there, above the Caribbean, was a cloud formation very similar to those from which devastating hurricanes often develop. Could it be that another disaster was brewing there?

For the next few days, the weathermen kept a close watch. Every time Essa sent back new pictures from orbit, they immediately looked for those phenomena which were so difficult to detect. It finally became evident: a new hurricane had developed. It was moving over the sea, small and insignificant; however, that was soon to change. Hour by hour "Camille" (as this hurricane was named) grew more and more powerful.

Camille sounds so human, like a nice little girl. The names meteorologists give to hurricanes are confusing, since these storms are not all that tender. On the contrary, they are furies. Perhaps the meteorologist who started the custom of giving girls' names to hurricanes had just had a fight with his wife, and maybe she reminded him of a raging hurricane. In any case, the custom was retained. Camille was to develop into one of the most violent of these furies.

The satellite report was then confirmed by observations from the earth. Ships gave a running commentary on

Camille's growth and on the cloud masses which were gradually spiralling upwards. Reconnaissance planes swarmed out, and flew into the center of the air disturbance to take additional measurements. The weather radar installations along the southeast coast of the United States were on alert. Everybody realized that what they were witnessing was only the "calm before the storm".

An army of American hurricane specialists was prepared for such situations. Regardless of how quietly Camille moved across the Caribbean, the hurricane would, inevitably, hit the continent, and then all hell would break loose. Nobody could say for sure what course Camille was going to take, but experience indicated that it would turn towards Mississippi and Louisiana. The population of this area was warned in mid-August, causing one of the greatest human migrations of our time. Seventy five thousand people left their homes and moved north.

During all this time, Camille remained quietly over the Gulf of Mexico. The storm was giving itself one last chance to catch its breath. Then, on the night of August 17, it struck. All of a sudden it came to life, swelled up, and raced faster and faster towards the U.S. coast. When it reached the mainland, its speed had grown to 195 miles per hour. The sea was whipped up, and the Mississippi overflowed its banks. Pass Christian registered a stormtide 26 feet above normal. The air pressure on the land sank to dizzying lows, as the barometer dropped to 26.6 inches. Only once, on August 18, 1927, had the air pressure ever been lower in the United States, and that was on the Pacific coast.

Nothing in Camille's path could stop her. Houses collapsed like papier-mâché: about 6,000 were totally destroyed, and about 50,000 were heavily damaged. Among these were houses which had been considered hurricane-proof for over 100 years, and which up to then had never been damaged. Their last hour had come.

What the hurricane itself spared was lost in the floods accompanying the storm. Property damage amounted to about

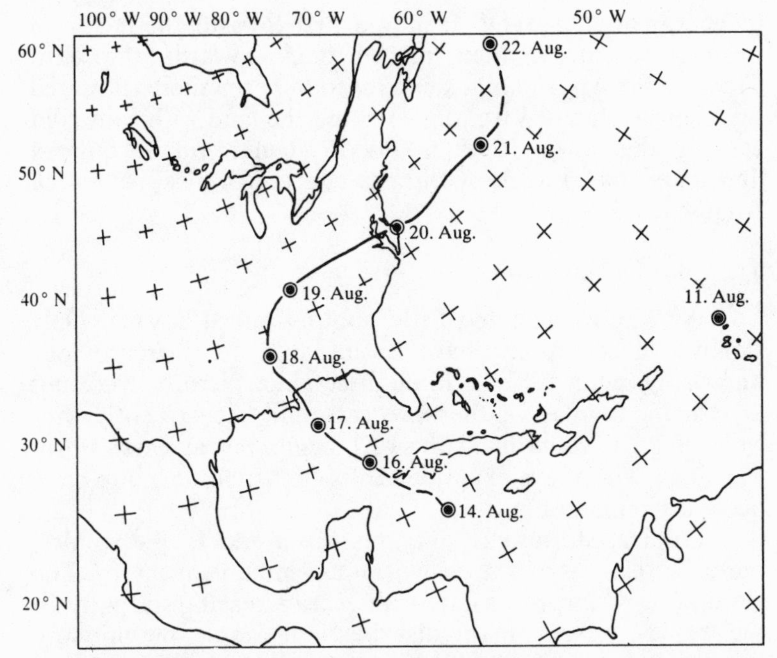

The path of destruction followed by Hurricane Camille. Thanks to early recognition and continuous surveillance from space, the amount of damage it caused was kept to a minimum. Untold thousands owed their life to satellite reconnaissance.

$1.5 billion, an enormous sum. In view of these figures, it is almost amazing that "only" about 260 people lost their lives. Never had such a strong hurricane raged through the United States. Thanks to the early warning from satellites many lives were saved. We can only guess how many people would have been killed without evacuation. The experts estimate almost 50,000, based on comparisons with other, weaker hurricanes, and their destruction. It is useless to wonder how profitable the weather satellites are; human lives cannot be estimated in financial terms.

Camille weakened relatively quickly. The hurricane lost

its energy over the land. It moved first towards the north, to western Kentucky, then turned back towards the east. Crossing the Appalachians weakened her even more. It reared up one last time in Virginia, covering the land with rain, and causing the James River to flood. Then it finally quieted down, and, as far as the world was concerned, disappeared on August 22nd.

The Life of a Hurricane

We know much too little about tropical storms. Only satellite observations have been able to increase our knowledge perceptibly. Although satellite pictures were not included in regular weather forecasts until fairly recently, they were of great value from the very beginning as early storm warnings. The above example shows that such warnings soon became irreplaceable.

Tropical storms are low pressure areas. In the tropics, these depressions can store enormous amounts of energy. The circulating air masses rotate with ever increasing strength. In the tropics, we can distinguish between simple low pressure areas (wind speeds up to 35 mph), tropical storms (wind speeds between 35 and 75 mph), and tropical hurricanes (wind speeds above 75 mph). Hurricanes, with diameters of up to 150 miles, are actually small in comparison to low pressure areas, that often cover 350–450 miles.

There are many names for hurricanes. They are actually called hurricanes only in the Atlantic and the lands that border on it. In the Pacific they are called typhoons. The Mauritius cyclones near Madagascar are the same thing, and in Australia they are called Willy-Willies.

For a long time almost nothing was known about these storms. They originate in the world's oceans, where, practically speaking, there are no meteorological stations. Until 1960 no one had ever been able to watch the "birth" of such a formation. Precisely for this reason, the inhabitants of areas threatened by hurricanes are pleased that there are now weather satellites to fill in the gaps in the observation network. Since 1962 almost all hurricanes have been detected by satellites.

This does not mean that before then people did not use all available means to discover hurricanes in time. Nobody likes to rely on chance. In 1956 the American Weather Bureau founded the National Hurricane Research Laboratory with its headquarters in Miami, Florida. American hurricane research is centrally directed from there. When called upon, ships and airplanes swarm out into the Gulf of Mexico to investigate suspicious cloud formations. Firmly anchored weather buoys and radar equipment along the coasts give additional information. All the reports are then sent to Miami, where they are evaluated by computer. When necessary, storm warnings are issued to the appropriate areas.

Nonetheless, the development of a hurricane during the first hours remained hidden until weather satellites started photographing the earth. In a picture which Tiros 1 took on April 19, 1960, we can see a small bright cloud system in the skies about 50 miles from Wichita Falls, Texas. It showed a strange square shape. In the beginning, no one knew what to make of this, but ensuing pictures revealed that what they had photographed was the birth of a hurricane.

In 1960 that was still something unique. The director of the satellite division of the American Weather Bureau, David Johnson, even spoke of a stroke of luck. Today we no longer have to say that. All we need to do is trace the course of a hurricane back to its origin on the weather satellite pictures. This they have now been doing for a long time. Anna, Debbie, and Esther are the names of the first hurricanes studied in this way.

Hurricane Anna proved to be an especially noteworthy phenomenon. On July 19, 1961, it was picked up by Tiros 3, and after this it was kept under continual observation. Gradually it approached South America. Its path led past Venezuela toward British Honduras, and there, to the surprise of the meteorologists, it broke up, for no apparent reason. When they traced its development back, they noticed that it had developed near Brazil.

As satellites observed more and more hurricanes, it became increasingly evident that what had until then only been surmised, was a fact: that breeding places for tropical hurricanes are in the world's oceans, precisely those areas inaccessible to conventional meteorological observation. In the beginning, these storms appear as small, insignificant cloud formations, which usually shine brightly. As soon as conditions are favorable, they rapidly change. An energy exchange of superdimensional proportions takes place.

Through some unknown mechanism, the cloud areas draw up the moist warm sea air from below. The clouds begin to rotate, leaving an "eye" in the center which is free of wind and clouds. At the edge of the eye the warm air rises. At higher altitudes it cools off, causing the air to become supersaturated with water vapor. This water vapor condenses, and the cloud formation grows larger and larger. The velocity of the winds constantly increases as a result of the heat energy released by this cooling process.

It is surprising how much energy the hurricanes or typhoons or Willy-Willies take up and store, even though their lifespan is seldom more than a week. These amounts are beyond comprehension. The energy exchange of a medium hurricane is equivalent to the amount of energy set loose when 1,000 atomic bombs are exploded per minute! It is no wonder that the destruction caused when this energy is released reaches such enormous proportions.

There are certain times when hurricanes are more apt to form. In the United States, the season runs from the middle of June until November. Within this period, the greatest activity extends from the middle of August until the end of September. These are the months during which the hurricane hunters are on maximum alert. Each season, about eight to ten hurricanes form in the western Atlantic, and we must expect at least two of them to strike the mainland, carrying out their destructive work.

It is not always possible to predict their path with 100% accuracy. Nobody can foresee accurately the life of each individual hurricane, even though experience has shown that they prefer definite "paths". Within certain limitations, early warning is always problematical. In 1964, the director of the weather bureau in Formosa found this out very tragically. He announced that hurricane Gloria, which had already been sighted, would not touch the island.

Gloria paid no attention to the prediction, and on the following day struck Formosa with all her might. The results were devastating: 239 people were killed, and 89 disappeared without a trace. The property damage remained within reasonable bounds – only about $18 million. The courts did not know how to handle this problem. The meteorologist responsible had done his best, but could not perform magic. He was thrown in jail for ten years on grounds of incompetence. Meteorologists live dangerously!

Similar things occur even in the United States, where the weathermen employ the most modern methods and the latest scientific advances to keep abreast of the course of hurricanes, and to evacuate threatened areas. In September 1965, for example, they could see clearly and distinctly on the weather satellite pictures that hurricane Betsy was leading a peaceful existence in the North Atlantic, and was giving no one reason for alarm. Then, all of a sudden, Betsy raced over Florida towards the Mississippi River, and, before the scientists understood what was going on, she struck. The insurance companies noted losses of one quarter of a billion dollars. This time the satellite pictures were unable to prevent a disaster. There is no way to protect against Mother Nature's moods.

On the other hand, one of the greatest natural catastrophes of our century could have been avoided, if the warning system in the rest of the world were as good as it is in the

United States. On November 12, 1970, the meteorologists in East Pakistan saw a dangerous cyclone on the most recent weather satellite pictures. It was approaching at a furious speed. Desperately, they tried to forewarn the population in the threatened areas, but without success. The country was so hopelessly overpopulated that their warnings got nowhere. The warning system was totally inadequate, and East Pakistan had to bear the consequences. Between 200,000 and 500,000 people lost their lives. The extent of the tragedy cannot be estimated more precisely but it may have been avoided by the utilization of satellites.

Establishing a Routine

Even though weather satellite pictures could not prevent catastrophes in these cases, it was nonetheless recognized from the very beginning how valuable they were for storm warnings in general. Tiros 1 made such important contributions to the early recognition of hurricanes that the idea arose of starting a systematic "hunt" for these scourges of mankind. Since Tiros 2 had already been planned for reconnaissance of ice conditions on the earth, its successor, Tiros 3, was designated for hurricane hunting. It was launched on July 12, 1961, right at the beginning of the new hurricane season.

At the same time a conventional hurricane warning service was set up. As soon as the pictures of the earth from Tiros 3 were delivered to the evaluation center, they were minutely analyzed to see if they exhibited any signs of a hurricane. Whenever such a formation was discovered, a warning was automatically sent to the threatened area. From the actual photographing to this point took no more than two hours. Although the investment was great, the success justified the means.

In the period from July 1961 to December 1964, over 1,000 storm warnings were sent to more than fifty countries in the world. While not all the storms were hurricanes, nonetheless 118 of them were caught! Naturally, some were also observed from earthbound stations, from ships and airplanes. In thirty seven cases, the Tiros satellites registered the hurricanes before they were observed from the earth. Because of this, early evacuation of the population was often possible. Probably thousands of people owe their lives to satellites.

Hurricane Carla, which hit the American continent in 1961, is a good example. Although it was one of the most dangerous hurricanes of recent years, it caused only very slight damage, because Tiros 3 had reported its existence to the earth three days before it began its destructive work. During these three days the threatened popoulation was evacuated. The people formed a heavy stream, as they moved deep inland in thousands of automobiles. They left the coast 300 miles behind them, and when Carla struck, only a few lives were lost.

In September, 1961, Tiros 3 was able to report hurricane Esther in advance of all other observation stations. Reconnaissance airplanes did not sight it until two days later. Again, the population had time to prepare for the storm's attack, and again the number of victims was kept to a fraction of what Esther might have claimed, had not weather satellites been used. Incidentally, Esther was also one of the first hurricanes whose strength the meteorologists tried to break artificially.

Tiros 3's successes were so enormous (experts talk about having avoided billions of dollars in losses, although such a statement can scarcely be proved), that the same method was used during the 1962 hurricane season. This time Tiros 5 was focused on the hurricanes. This is the satellite which earned itself the nickname "hurricane hunter".

Actually the nickname is not very exact, since Tiros 5 surveyed not only the Atlantic, but all tropical areas of the world as well. For example, in August 1962, it succeeded in spotting a storm in the Indian Ocean which up to this point had gone undetected. On the basis of Tiros 5's pictures, the people living on Africa's east coast were made aware of the threatening danger. Once more a satellite had proved itself.

In the long run, however, people were not completely satisfied with just storm warnings. It was true that many losses were avoided, but often even the satellites did not detect hurricanes until too late, because these storms are often not recognizable in their early stages. In addition to satellite observations, it was necessary to employ hurricane reconnaissance airplanes, and this was an expensive venture. These flights could only be restricted if the meteorologists were routinely successful in assuring hurricane warnings by means of satellites. That meant hurricanes had to be studied even more closely.

In the early stages of weather satellite photography, the scientists had already recognized that every hurricane had a characteristic form. They were all individuals, to be sure, but several showed a certain familiar likeness. Amazingly enough, the meteorologists could already tell from the form of the storms whether they had arisen in the northern or southern hemisphere. North of the equator they usually have very circular spirals, while south of the equator, they are generally stretched out along the meridian. The form was thus influenced by certain regularities in development.

The reason for this north-south asymmetry was soon discovered. It was a question of heat and energy transport. In the northern hemisphere, the large mountain ranges such as the Himalayas, the Rockies, and the Alps, as well as the continents themselves, have a great influence on this exchange. In the southern hemisphere there are neither the large mountains nor a sufficiently large land mass, so that here hurricanes are primarily responsible for the exchange. For this reason, south of the equator there are a greater number, and more intensive hurricanes than in the north. This fact had gone unnoticed until then, simply because there are too few observation stations in the southern hemisphere.

In the spring of 1964 they started looking at such differences in the appearance of hurricanes. Hundreds of satellite pictures were studied closely. They even studied the sizes of the formations, and these factors were simultaneously compared with the maximum wind speeds registered. At first there did not seem to be any correlation, but then it turned out that at least the maximum wind speeds depended on the hurricane's size and structure.

This correlation was tested during the hurricane season in 1964. As soon as a hurricane was discovered, the scientists predicted the maximum wind speed that would be developed. At the same time, the reconnaissance airplanes swarmed out to measure this speed on the spot. Lo and behold, there was an agreement. On the average, the mistakes in prediction ran about 10%.

The most fantastic prediction was that made in the case of hurricane Cleo. Based on a weather satellite picture of August 26, 1964, the meteorologists estimated Cleo's maximum wind speed at 73 mph. The reconnaissance planes then reported an actual value of exactly 73 mph. Thus they reached yet another goal in satellite meteorology. Now it was possible to predict the exact life pattern of a hurricane. Consequently, the expensive airplane reconnaissance flights were decreased. The Tiros satellites had performed their first important routine job.

A Large-Scale Attack on Nature

Man is not born to flee from danger. When he is threatened, he tries to remove the source of the threat. This is not always easy, especially when the threat is in the form of a natural catastrophe. Here the battle is between unequal forces: nature is incomparably stronger than man. Nevertheless, meteorologists were bound to start wondering whether they could not nip the power of hurricanes in the bud through early recognition. In the long run, early warning by itself was simply not sufficient. It did help keep the damage caused by hurricanes to a minimum, but could not prevent it entirely.

A new idea was close at hand. It was inspired by the successes (and even failures) of "rainmakers", who had been causing a stir for decades. They had first surfaced at the turn of the century, especially in the United States and Australia, moving through dry regions and claiming they could make the clouds rain. They knew how to appear scientific, and people were usually gullible enough to believe them. Maybe it was nothing more than mere hope that moved people to pay these rainmakers.

They were proven charlatans and adventurers who took advantage of the limitless trust of the rural population to pursue their illicit handiwork. They did not worry at all about their results, although occasionally they did have considerable success. Good luck seemed to favor them when they conjured the clouds, and the farmers who finally saw rain again, after weeks of drought, believed in the marvelous gifts of the rainmakers.

Today we are enlightened enough to see through these adventurers. Nonetheless, we have learned from them. Scientists have taken over the rainmaker's job. They have had some successes which are so impressive, that the dream of controlling the weather has been partially fulfilled, even though it is still doubtful today that complete control will ever be possible. Maybe we should hope that this will never come about, because rainmaking leads to problems which could easily develop into conflict. What would happen if the precious clouds were tapped over one piece of land, while a neighbor maintained that the rain belonged to him, since, if left alone, the clouds would have burst over his property?

Scientists were not at all concerned with these problems when, in the first half of our century, they started investigating the causes of precipitation. After lengthy research they found two answers, both of which presuppose the growth of precipitation particles in the clouds. In one case, these particles grow when smaller particles flow together. Beyond a certain point, the buoyancy of the air can no longer support them, and they fall to the earth as precipitation.

The second answer assumes the existence of ice crystals in the clouds. Because these icy crystals have lower temperatures, the water vapor saturation pressure near them is lower than near water drops. Through a simple physical process, which we will not discuss in detail here, the ice crystals now gather water drops from their surroundings. Again, the weight of the individual particles increases, and this then results in precipitation.

The scientists had thus identified the causes of precipitation. A method to produce rain artificially was at hand. They only needed to sprinkle the clouds with ice – for example in the form of dry ice – and precipitation would follow. This is actually possible. However, scientists have developed a different method which is more popular. They recognized that the crystalline structure of silver iodide is very similar to that of ice crystals. Thus it is also possible to "seed" a cloud with silver iodide. Water drops will gather on it in the same way, and rain will fall. Of all the methods for tapping clouds, this is the cheapest and surest.

The method is somewhat more complicated in practice, and is not always feasible, but it has often led to success. In the mid-fifties, the first steps were taken towards commercial use of cloud seeding. Within certain limits rain could be controlled, and the damage caused by a long-lasting drought could often be avoided. On the other hand, over-wet areas were also spared additional large amounts of rain, because the clouds were tapped before they reached these areas. The meteorologists also succeeded in artificially weakening hail storms, and in dispersing fog. Meteorology had become a practical science. Hurricanes, however, possessed such a great amount of energy that taming them was not possible at first, and they remained untouched by these experiments.

Soon, encouraging signs began to appear. In theory, it is possible to master hurricanes. A large portion of a hurricane's energy is stored as heat in its water vapor. One should be able to release this energy over uninhabited areas before the hurricane can cause destruction through its motion or, primarily, because of the distribution of air pressure which is typical of hurricanes. For this purpose too, early recognition of hurricanes by satellite is especially important. Attempts to defuse hurricanes must be carried out at a stage when the formations are still located over the ocean. The danger of an unwanted disaster would otherwise be too great.

A decade ago Robert Simpson expanded on the basic idea that a significant amount of heat energy is released by the condensation of water vapor and by the formation of ice crystals. He suggested that silver iodide crystals offered the key to success. A subdivision of the American Environmental Survey Sciences Administration (Essa) responsible for physics and chemistry set to work on the problem. They founded Project "Storm Fury" whose objective was, among other things, to determine whether hurricanes could be broken up before they became destructive.

They carried out their first large-scale experiments under Project Storm Fury in 1961. As we have already mentioned, this was the same year that the weather satellite Tiros 3 located hurricane Esther long before it was spotted by ground stations. Immediately, telephones started ringing as the necessary offices were informed. After a short hectic period of preparation, several airplanes took off to find hurricane Esther.

Directly below the storm, the order to fire was given. Eight 145 pound bombs equipped with steering fins left their racks. For once, these were peaceful bombs. The airplanes turned back, and the bombs released their contents: silver iodide. A warm air current seized the crystals, and shot them high up into the air. Here they settled down as foreign bodies inside the hurricane.

One hour later radar stations watched the results. Esther's water vapor had partially condensed, forming ice crystals, and fell to the earth as precipitation. Hundreds of cubic miles of rain disappeared into the Atlantic, without having caused any damage. The amount of heat released was considerable. According to some estimates, it was equal to the energy of eight 20,000 ton atomic bombs! Esther's wind speed was reduced by 14%. This was the first achievement in hurricane seeding.

The scientists involved in the project expressed only muted optimism. They could be proud of their work, it is true, proud that they had reached a milestone, but they were well aware that their success was only temporary. Two hours later the hurricane had already recovered; it had regained its original energy and wind speed.

We must not let that overshadow the fact that the experiment was a success. After all, the whole undertaking was the very first scientific attempt. Scientists had to make new, broader studies before they could better understand hurricanes and their dynamics. In 1963, additional hurricane seedings were made. Again and again they proved that the energy released was virtually unmeasurable. Any partial successes gained were soon negated, however, because the storms quickly recovered. Things are still the same today.

There was only one way out of this dilemma: the hurricanes had to be attacked while they were still developing. The scientists knew the essential pattern of this development from their studies of weather satellite pictures. Thus, attacking the storms at an early stage should not have caused any difficulties. The first favorable opportunity came in 1965. One of the weather satellites shot a picture over the Caribbean which showed nineteen hurricane-like formations.

Essa's reconnaissance planes immediately swarmed out, and within a short time they had verified the existence of the clouds. Further airplanes were now loaded to the bursting point with silver iodide bombs, and another large-scale experiment began. Twelve of the nineteen cloud formations were seeded. The rest were left as controls, and were allowed to develop freely under continual observation.

The experiment was a complete success. The non-seeded

clouds followed the typical development of tropical storms. Although they did grow, their growth remained within bounds. (Let us remember that the diameter of a normal low pressure area is generally far greater than that of a hurricane.) The seeded clouds, on the other hand, proceeded to expand more and more. Some of them even reached enormous heights. They hung in the air weighty and powerful, but harmless. Some of them finally produced heavy rain storms. The seeding had apparently been successful.

In the next few years the meteorologists continued to treat hurricanes in the same way. In 1967 a long-range experiment was undertaken which lasted more than two months. From August 8 to October 15, fourteen American Navy planes stood ready to fly out and drop their silver iodide bombs. The seeding actions in September are noteworthy. The pictures which the weather satellite Essa 5 took on September 14 show three clearly recognizable hurricanes: Beulah, Chloe and Doria. Another hurricane was still in its early stages. Beulah, Chloe and Doria were treated. The hurricanes did weaken, but, again, they soon regained their original force.

A few days later Doria and Beulah moved towards the American continent. Beulah especially was raging furiously at this stage. It moved westward from the Caribbean, crossed a part of the Yucatan coast, and then moved towards the northeast, causing great destruction in Texas on September 20th. Then it turned back toward Mexico once more, where its energy was finally used up.

Another interesting experiment in hurricane seeding was carried out in 1969. This time the "patient" was named Debbie, and was treated on August 18th. American Navy planes flew over it several times and covered it with a twenty mile long silver iodide carpet. A few hours later the storm had weakened. Its wind speed had decreased from 110 mph to 76 mph. On the next day, Hurricane Debbie recovered from its shock. Yet the scientists refused to leave it in peace. On August 20 they tried a second seeding and, again, the wind speed sank by 15%. However, this success did not last either.

All of these experiments led to one conclusion: with modern methods we can reduce the strength of hurricanes, but the results are not permanent. Total success is only possible if the storms are treated in their initial stages. For this, use of weather satellite pictures in early recognition is absolutely essential, since the earthbound meteorological observation stations are generally not in a position to sight hurricanes while they are still in this stage.

In spite of these achievements, not all scientists have admitted the value of seeding hurricanes. It is very difficult to decide whether a decrease might not have come about by itself, without seeding. In many cases there are not enough data to clarify this question. In 1969 the World Organization of Meteorologists published a balance sheet which must have been very depressing for the scientists. Of the twenty three seedings that were carried out, only six were so successful that more rain than normal was registered as a result. As a matter of fact, in ten cases less rain was measured. The rest of the seedings led to no definite results.

Only twenty three tests are cited in this report because accurate statements can only be made after very careful investigation. Many weather makers do not follow this principle, so that J.B. Mason, Director of the British Weather Bureau, recently felt called upon to take a public position against all experiments to influence the weather. Since the end of 1971 it has been mandatory in the United States to report any such experiments, including hurricane seedings. This is to assure the proper preparation and careful execution of future experiments.

Because of certain additional data, the treatment of hurricanes is no longer considered as pessimistically as in the 1969 balance sheet. Although the achievements of the rainmakers were challenged, the picture as far as they are concerned has now changed somewhat. In the United States computer calculations helped prove that there are a series of cloud types which are especially susceptible to silver iodide. With these particular cloud types, rain could be increased through seeding by 140% on the average. the possibility that this is pure chance is one in two hundred!

Perhaps one day we will finally be able to master hurricanes. A question we must ask ourselves is whether we want to reach this goal. After all, hurricanes, especially in the southern hemisphere, play an important role in the exchange of heat energy. At present nobody can say what effect it would have on the world's climate, especially in the non-tropic lands, if this heat exchange were to be eliminated by artificially destroying hurricanes.

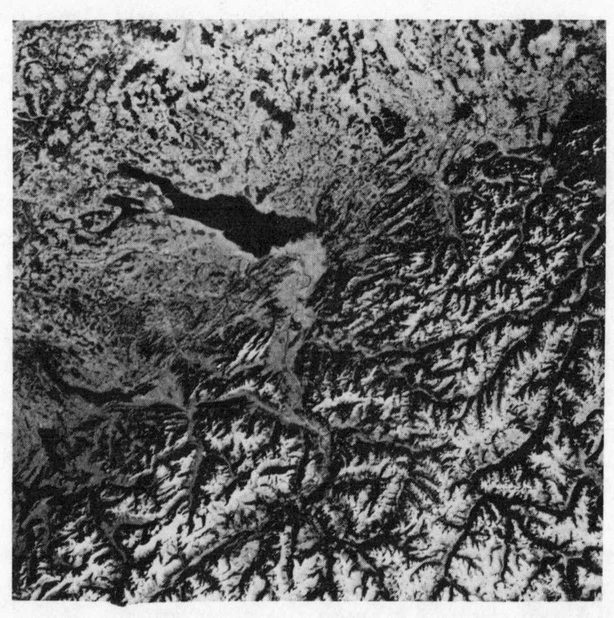

Multispectral photo of Lake Constance and surroundings, taken at 570 miles. Even on this black and white copy of the picture, the larger cities stand out as dark spots.

8. Organization of the World Weather Watch

Automatic Picture Relays

In December 1966 the inhabitants of the Fiji Islands discovered just how valuable satellites could be. Essa 3 had spotted another hurricane, which this time was moving straight towards the charming island group in the Pacific. The population was immediately warned of the approaching danger. Express telegrams were sent to the neighboring Yasawa islands, and air traffic was shut down. When the hurricane finally moved through the islands, it did not cause any noteworthy damage. For the first time, the American satellite storm warning system had proved its value in the Pacific.

The natives of the Fiji Islands learned from this experience. They bought their own ground stations, so that they could directly receive the weather satellite pictures taken over their area. Thus they followed the example of many other countries, which had had nothing but good results with such stations.

Direct reception of weather satellite pictures has been possible since 1964. With the exception of the Tiros storm warning service, until then it took a very precious six to eight hours before weather pictures from the Tiros satellites reached their destination. This delay made it impossible to use these pictures in weather forecasts. The problem haunted the Ameicans until finally they found a solution. The launch of Tiros 8 on December 21, 1963 enabled them to test APT (Automatic Picture Transmission), a necessary feature for a worldwide weather observation network to be used for other than pure research.

The APT system was not completely functional until the launch of Nimbus 1 on August 28, 1964. NASA was convinced that the Tiros satellite system would be ready for regular use in a short time. There would always be at least two weather satellites orbiting the earth, continuously relaying pictures of weather events. Since NASA itself was concerned with research, and not with routine duties, it wanted to hand supervision of the satellites over to the U.S. Weather Bureau. NASA hoped to facilitate this by introducing the Tiros Operations Satellites (TOS), whose predecessors were Tiros 9 and Tiros 10. However, they also had to test modernized weather observation techniques so that they too could be included in the routine jobs. This was the purpose of the Nimbus satellites.

The Nimbus satellites proved to have been well conceived. In fact they are still in use today, and form the basis for the earth reconnaissance satellites. The secret behind their success is the "Drawer Method". Basically, all Nimbus satellites are extraordinarily similar; only the scientific equipment is different from flight to flight. NASA took this into account by providing each satellite with a standardized observation platform on which equipment could be exchanged without any problem, just like taking out one drawer and putting in another. This did not effect the basic structure of the satellites, which could thus be used for many different purposes.

Naturally the primary function of the Nimbus satellites was to test methods of observation. For this purpose their stabilization was ideal: their observation platforms were always directed at the earth. This had been a very complicated achievement. Sensors sought the sun, and registered the position in space; based on this, a computer calculated the corrections necessary to maintain the proper attitude toward the earth.

In addition, the "sun paddles", the satellites' outriggers

containing thousands of solar cells, always had to point toward the sun in order to transform sunlight into electrical current. That meant that these sun paddles were constantly in motion with respect to the rest of the satellite. Consequently, the Nimbus satellites were mechanically very complex and this, in addition to having a computer, made them quite vulnerable. For this reason, their use for routine work was not at first considered. Nimbus 1, for example, ceased functioning on September 23, 1964. Nonetheless, it had relayed more pictures of the earth during its short lifespan than Tiros 1 had in three months.

Nimbus 1 was thus used to test the APT system. This system was based on a simple premise: the faster a picture is relayed, the more noticeable are the disturbances in the relay. The errors caused by static can only be decreased by using expensive receivers, although even then they cannot be eliminated entirely. It is difficult for poorer countries to purchase these receivers, since they cost several million dollars. The solution to the problem was therefore to relay the pictures more slowly.

This was not easy, since the pictures of the earth which the television cameras took first appeared on a picture tube, and it was these pictures which were then scanned and relayed to the ground stations. The length of time that the picture was on the screen had first to be lengthened; that was the function of the newly built APT cameras. The cost of building a reception station was thus suddenly diminished. The expensive television receiver was replaced by a facsimile picture receiver, which anybody could afford. The results of this development can scarcely be estimated. Not only do the governments of many countries now possess weather satellite receivers, but also private firms, and indeed private individuals. At the beginning of 1972, 94 countries possessed a total of 550 APT stations, with new stations being added every month. In the United States there is no problem in acquiring such complete stations. They do not have to be specially constructed: all that is needed is $10,000. For this money anyone can participate in space exploration. Weather satellite technology disproves the claim that space exploration costs billions of dollars, and that therefore only rich nations can afford it. In fact, the American APT stations are not even the cheapest.

The French have undersold the Americans. The French Post Office uses mobile reception stations weighing no more than 220 pounds. They are produced in the French Research Center for Communications Technology at Lannion. People have gradually come to realize that the "expensive" American machines are not absolutely essential for receiving weather satellite pictures, although, understandably enough, the Americans still have the largest share of the market.

The receivers are basically so primitive that there is no reason why an individual could not build one himself. Astonishing as this may seem, a group of students in Montgomery County, Maryland is constructing satellite picture receivers in their own homes. Wendell Anderson, a ham radio operator from Moorestown, New Jersey, spent very little for his station.

Anderson receives weather satellite signals with a ordinary amateur radio, and then records them on a tape recorder. In the second state of his work, he uses the tape recorder to regulate the brightness of a lamp. Aimed through a microscope, the brightness produces dots of light on ordinary photographic paper, which itself is fixed on a rotating cylinder turned by an electric motor. A second electric motor pushes the microscope from left to right, so that a complete weather satellite picture evolves line by line. The whole station costs less than $600!

Amateurs at the Walter Hohmann Observatory in Essen, Germany have been receiving weather satellite pictures for quite sometime. They use a facsimile machine which the press gave them when they no longer needed it. The rest of their station they jerrybuilt for only $250. They were even able to receive broadcasts from the more recent NOAA satellites, which says a lot for the quality of their station.

With the APT system, it took considerably less time for national weather stations to get cloud pictures. On the other hand, these pictures were no longer magnetically stored. Basic meteorological research would have come to a halt, if the APT system alone had been employed. To collect the photographs of single regions of the earth at one central location would have been enormously expensive. For this reason, in addition to its APT camera, Nimbus 1 carried another system of three cameras similar to the photographic equipment used in the Tiros missions. Pictures from these cameras were again stored on magnetic tape, from which they could be recalled by two CDA stations in America.

Although the photographic material rapidly increased, evaluating it was much easier than for the pictures from the Tiros satellites. Since the Nimbus satellites were always aimed at the earth, there was no longer any distortion in the pictures due to improper aiming of the cameras. Those distortions occasioned by the shape of the earth and by the camera optics themselves were always the same, so that even before the pictures were taken, grids containing the geographic coordinates could be made for the photographs. Meteorologists could thus lay over the grids and begin their analysis as soon as the pictures were received, gaining valuable time. Nimbus 1 had opened up new dimensions in satellite meteorology.

Dividends from Weather Satellites

Over the course of years, weather forecasting has gradually improved. Although many people do not believe this, statistics prove it beyond a doubt. We need only check the weathermen's predictions whether it will rain or not, and we will have to come to this conclusion. Eighty seven percent of the time this simple forecast is accurate for the same day, and, for the following day, eighty percent of the time. This means that weather prediction over a two-day period is now better than it was for only one day when Tiros 1 was launched.

It is obvious that improving weather forecasting has

meant important gains for all sectors of the economy dependent on the weather. Not only agriculture, but also the food industry, construction, air and ship traffic, energy production, and water supply all profit from it. While we cannot fully determine to what extent weather satellite meteorology has already paid for itself, that it has done so is indisputable.

The uses of nephanalyses alone show this. Weather pictures from Tiros 9, obtained twice daily from the U.S., enabled the British Weather Service to avoid making an inaccurate forecast on February 8, 1965. Such an error in forecasting would otherwise have been unavoidable, and most of the population would have understood. Weather conditions change so rapidly and so unexpectedly in this part of the world that the meteorologists are faced with enormous problems.

Such was the case on February 8th. All indications pointed to bad weather, as temperatures in Great Britain sank to 35-37° F along a widespread, turbulent coldfront accompanied by showers. Reports from separate weather stations were in remarkable agreement. Radio probes into the atmosphere from Hemsby, Norfolk reported rain. A ship crossing the North Sea at the time also reported cloudy skies. There were no signs of clearing weather anywhere.

At the last minute the British received a nephanalysis from the United States which changed the picture completely. It showed a broad zone of clear sky over the North Sea. The meteorologists could thus quickly change their forecast, and so hit the bull's eye. Their revised forecast was entirely correct. Dry air moved over the island kingdom in the evening, and during the night it turned clear and frosty.

Toward the end of 1960, Australian meteorologists also profited from satellite pictures. A long heatwave had lain over Australia, and it seemed that it would never end. The farmers were getting prepared for more hot days. Then the weather forecasts had to be changed. The Tiros pictures showed a broad coldfront approaching from the Antarctic. Because of the scarcity of observation stations south of Australia, the change in weather had not been picked up by ordinary methods. Within two days the coldfront reached the continent, proving again the value of satellite pictures.

The Australians have attempted to estimate how much their economy owes to weather satellites. Their careful figures reached a value of 50 million Australian dollars per year. Even though we must look at this estimate cautiously, it is probably not too far off the mark. The French have also estimated that they increased their grape and fruit harvests so tremendously in the last few years that they now earn about 200 million francs more per year. In Germany the profit due to weather satellites, figured over the entire economy, probably reaches one billion marks per year. These estimates, which should also be viewed with caution, are based on extensive polls. For statistical purposes, branches of the economy were polled which were dependent on the weather. Each of them gave their opinion on a number of questions, resulting data were considered objective.

In spite of very conscientious investigations, there are still a number of uncertainties in these evaluations. Thus it is completely understandable that the publicized figures are strongly divergent. This becomes particularly clear, for example, when we look at the financial benefit which satellite meteorology is supposed to produce in Germany, and compare it with the conclusions of the Congressional Committee for Science and Space Exploration a few years ago. The committee's report stated: "Improving the accuracy of weather forecasting by only 10% would mean savings of several hundred million dollars in agriculture, construction, air and water shipping, and many other businesses." We cannot determine whether the German evaluation was set too high, or the American too low, but we should bear in mind that the U.S. is much larger than Germany.

That does not alter the general importance of satellite meteorology. Indeed, in special cases, observations in greater detail could be made. As an example, we could cite a recent study made in the U.S. concerning the hay harvest. This study showed that hay must dry for three days after it is cut. If it rains during this period, and the hay gets damp, the protein content decreases quickly, which means a substantial decrease in the value of the hay.

If a dry period of at least three days could be predicted when the hay is ready for cutting, then the date of mowing could be set accordingly. The hay would maintain a maximum protein content and its value would increase so much that the the profits would rise by about $28 million per harvest, per state. Perhaps in a few years satellite meteorology will have progressed to the point that it can make this dream come true.

A reliable extended forecast is still not possible, but scientists are working on it. They are trying to expand their global observation system so that they will have sufficient data to predict the weather over a fourteen-day period, using complicated mathematical models. Farmers will then be able to do their sowing and harvesting at a time when the temperature and precipitation are most favorable.

Agriculture is always cited first as the branch of the economy which benefits most from weather satellite technology. It is however only one of the many branches. Also, it is often not information about the weather itself that is necessary, but about all kinds of other phenomena which could lead to disaster and misfortunes.

For example, in the summer of 1963, Tiros 6's pictures showed a strong sandstorm in Saudi Arabia. No other observation post had noticed it; there was no reference to it anywhere. Naturally the country's oil industry was extremely glad when it got a cable of warning from the United States. No one would otherwise have taken any precautionary measures against the storm, and who knows what the consequences might have been.

We can say with some assurance that the Mexican cities of Gomez-Palacios and Torreon, would no longer exist, if it had not been for a warning from weather satellite Essa 6. The authorities had failed to observe that a nearby reservoir and adjacent streams were approaching a flood level. Sooner or later this would inevitably have led to a catastrophe. Essa 6 stepped in with a helping hand. Its pictures alarmed the authorities; the necessary measures were taken, and disaster was prevented at the last minute.

Avoiding disaster is not the only profit to be gained. This became clear in the Antarctic, in a far-reaching program in which the continent, for the most part completely unexplored, was to be surveyed and mapped from the air. The program was both costly and time consuming. Pictures of the earth from the weather satellites helped in predicting cloud-free days for the mapping flights, so that almost no flight returned empty handed. This could save about $100,000 in a single season.

Safety in Air and Ocean Travel

Weather satellite pictures have become almost indispensable for air travel. They perform valuable services everywhere, but especially on flight routes through areas which had previously had almost no weather reports, because they lacked weather stations. For example, flights from San Francisco to Hawaii, or from Hawaii to the Fiji Islands, cross the open seas where ships do not ordinarily go. Until the launch of Tiros 1, this area was a no man's land for meteorologists.

Even where traditional observation stations exist, they are not adequate for a complete survey. Many meteorological phenomena go undiscovered. Usually just one look at a weather satellite photo is sufficient to localize these phenomena, and to enable us to give forecasts which are priceless for air travel. The reason is that a number of meteorological happenings are connected with very characteristic cloud formations. These formations only need to be spotted in order to predict the weather.

An example of this is the tropical hurricane, which we

have discussed in detail. This type of storm has already been adequately classified. No less remarkable are the pictures of the famous "jet streams" – those high altitude winds which are always accompanied by long, extended cloudfields. They range in speed from 125 to 375 mph. An airplane which rides one of these streams can easily save significant quantities of fuel, not to mention time.

This has been demonstrated repeatedly. The first flight competition which made the phenomenon of the jet stream public took place on May 2, 1952. An airplane started from Labrador to London. It flew on a direct course towards the British metropolis, so that it could not possibly have been passed by another airplane of the same type under the same conditions.

The BOAC pilot, Bernhard Frost, proved that meteorological conditions must also be included. He took a leisurely look at the latest weather maps, and noted a jet stream over the North Atlantic. He took off for London fifteen minutes after the first airplane. However, he did not fly directly, but rather turned towards the north to take advantage of this fast, high altitude wind. In so doing, he made a detour of 170 miles. He landed in London, having used less fuel, twenty minutes before the first airplane.

Ever since weather satellite pictures have been routinely used in air travel, such cases are no longer remarkable. These pictures give such a clear survey of the weather that even today, many pilots still react enthusiastically. The report of one pilot is well known. On a flight from New York to Dakar, he was able to rely on Tiros pictures, and his claim does not stand alone: "From 60° west of Dakar on, the map was so exact that it was almost unbelievable. Not only was the coldfront precisely located on the map, but, within the 150 mile scope of our airplane radar, its northeast – southwest course could be clearly recognized. The cloud types and frequencies were exactly reproduced. . . I think we have found the answer. Let's launch more Tiros satellites."

By now, weather predictions for aircraft are indeed

based on reports from satellites. In 1962, Kennedy International Airport set the example by having the latest nephanalyses sent to them regularly. Other large airports in the U.S. soon followed suit. Subsequently they also installed APT stations. Even in Germany people could not close their eyes to the new development. On July 15, 1966, the German Weather Bureau established direct picture contact between its APT station in Offenbach and the Flight Weather Watch at the Rhein-Main Airport in Frankfurt. Pilots have now become much more aware of the weather along their flight path, and as a result, air travel has become decidedly safer.

One can definitely predict that this is not the final development. Some pilots are already looking for the installation of APT stations on board their planes, so that they can get the most up to date information during their flight. On the other hand, they would even be satisfied if they could get the information directly from the airports. This is why they urge the establishment of a modern traffic supervision system based on satellite reports, instead of a satellite navigation system alone. This is understandable.

What we have just discussed concerning air travel is in general also true for shipping. In this case however, it is much more difficult to clearly prove the value of weather satellite pictures. Weather conditions on the sea change much too quickly to make predictions which would be relevant for more than one ship.

The voyage of the Soviet oceanographic ship "Professor Vitse" is an interesting case. In December 1967 this ship set out into the Antarctic to take various measurements along the route to the Mirnyi Observatory. There would have been no problems if the ship had not been completely incapable of crossing thick ice fields.

Although December is midsummer in the Antarctic, that does not mean very much. Even then ships must contend with icebergs and ice floes. Although a real possibility of danger threatened "Professor Vitse", the ship started out anyway. The Soviets had provided it with the latest ice information gathered from pictures taken by their weather satellite Cosmos 184.

At the beginning of the voyage, they drew a map which included the ice conditions near the Mirnyi Observatory, based on pictures Cosmos 184 had shot from orbit. The map also included information about the general behavior of ice in the Antarctic. Earthbound weather stations were almost nonexistent in this area. The best route for the research ship was then determined.

On December 10, after receiving a new photograph from Cosmos 184, they revised the map and with additional photos from December 15–17, brought it up to date for the last time. Success was not long in coming. "Professor Vitse" discovered one isolated iceberg on December 17, and one large ice floe on December 20. The ship reached its goal safely, but without the weather satellites, the trip would probably never have been undertaken.

Heat Pictures of the Earth

The value of photographing the earth from weather satellites was recognized in 1960, soon after the launching if Tiros 1. It was not long after this that the idea arose of setting up a global weather satellite system which would work around the clock. This plan was based on the experiences gained from the Tiros satellites, which after all had only be planned as research satellites. In 1961, during the presidency of John F. Kennedy, Congress authorized the National Operational Meteorological Satellite System, to be under the direction of the U.S. Weather Bureau. This is a branch of the Environmental Survey Sciences Administration (Essa), which in turn is part of the Department of Commerce.

The APT system was an essential prerequisite for creating this system. Another was night photography of the earth. Since the weather is always changing, limiting observation of the earth to daytime only can be but a stopgap measure.

It took several years, however, for a suitable night camera to be developed and perfected to the point that it could be tested in actual use. From the available choices, the meteorologists selected a type of night photography based on the principle that every body, including the earth, possesses a certain inherent temperature. According to this temperature, it emits radiation, especially in the longwave infrared ranges of the electromagnetic spectrum. This radiation can be recorded, giving "infrared pictures" of the body. For example, it is well known that large bodies of water give off more warmth during the night than do land masses, and that clouds moving high above the earth are appreciably colder than lower clouds. This means that if we examine the earth at night with infrared detectors, and use these measurements to direct a light source so that a strong infrared radiation causes a weak light impulse and a weak infrared radiation causes a strong light impulse, then pictures can be produced which are similar to real photographs: i.e. they show clouds bright, land dark, and bodies of water even darker. (During the day the land on infrared pictures often seems darker than the sea, because the sun heats the land and sea at different rates.) This is the principle employed by the infrared cameras in modern weather satellites. These cameras, like the APT cameras, first proved functional on Nimbus 1. They scan the earth line by line, and because of the satellite's own motion, gradually produce a picture of the entire area flown over. In practice, such a picture covers a continuous strip of land around the entire earth. It develops piece by piece, and there is no reason to break off the process at any one point.

The infrared cameras could not be tested until Nimbus 1, because their detector at first possessed an extremely poor directional resolution power. Consequently, the pictures were so coarse that precise details were unrecognizable. By 1964, although the resolution of the infrared pictures was not as good as that of actual photography, it was good enough to be tested.

Infrared pictures had another special advantage. Since

they were actually pictures of warmth, they could be used for research impossible with normal photography. They have proven especially useful in observing the Gulf Stream and other warm or cold ocean currents. The Gulf Stream plays such an important role in climatology that research ships have been charting its course and variations for years. This process was extremely time consuming, since the ships had to take many temperature readings throughout almost the entire Atlantic. On infrared pictures the warm Gulf Stream is clearly differentiated from its colder surroundings; one single picture can replace weeks of research trips.

Furthermore, the infrared pictures are essential to the investigation of the polar regions. They show the boundaries of the Antarctic continent with great precision. These boundaries are never visible because of the layers of ice which cover them and stretch far into the ocean. The thickness of the ice layers can also be estimated. From this, too, we have been able to increase our knowledge of climatology.

The military are probably the main users of the infrared pictures. They have built their own cameras into espionage satellites to take pictures of our earth day and night. The cameras are so accurate that they immediately register any variation in the earth's radiated warmth. These fluctuations often lead to vulnerable conclusions.

A few years ago, the Americans surprised the world with a 180° shift in their policies. They no longer, as previously, insisted upon on-the-spot examinations under their nuclear arms pact with the Russians. This change in attitude is not really surprising, however. Where people work with nuclear fission, large amounts of water are necessary to cool the equipment. The water is usually taken from rivers, and then returned at a much greater temperature. A rise in river temperature shows up clearly in infrared pictures. Space exploration has thus made possible the close inspection called for under the nuclear arms agreement.

With the help of infrared cameras, the espionage satellites are also available to track atomic submarines, whose equipment likewise warms up the surrounding water. It is believed that the Soviets are the ones who take advantage of this fact, since the Americans in referring to this technique have indicated that their atomic submarines are no longer as invulnerable as they originally seemed to be.

In spite of the impressive results obtained with infrared cameras, the first Tiros operational satellites had to make do without them. Apparently the scientists wanted to collect additional data without giving up their operational satellites whose life span was gradually coming to an end. In 1966 they inaugurated the new global observation system. Its satellites, which we have mentioned several times, were named Essa, since they were under the control of the Environmental Survey Sciences Administration.

The basic structure of these satellites, especially as regards their stabilization in space, was very similar to the "wheel satellites" Tiros 9 and 10. They "rolled" across the earth at an altitude of about 875 miles, and always shot their photographs when their cameras were pointing perpendicularly to the earth. This method had proven to be the most suitable. The stabilization technology of the Nimbus satellites was still too costly for an operational system.

Their camera technology relied on the experience gained from the Nimbus satellites. The Essa system used not only APT cameras, but also cameras which stored pictures on magnetic tape ready for retrieval by the CDA stations in the United States. Because of the high altitude, the magnetic tape was able to store enough pictures from one complete circuit of the world. For technical reasons the tasks were divided. Two satellites circled the earth simultaneously: one used the APT system, and the other the second system. Whenever one of these satellites failed it was immediately replaced by another.

Essa was launched on February 3, 1966. It was the first in that series of Essa satellites equipped with two cameras whose pictures were stored. During each orbit, the pictures were retrieved and forwarded to an evaluation center. There, with the help of a complicated computer program, they were combined to form three earth projections. This was a major innovation. One projection was of regions near the equator, while two were of the polar regions. Since the sun shines on every area of the earth once daily, a complete picture of the earth could be broadcast to the whole world. Those areas at the polar caps which were in the winter half of the earth, were all that remained black on the pictures.

The other group of Essa satellites, to which Essa 2 belonged, lifted off on February 28, 1966. These had two APT cameras whose pictures could be received directly anywhere in the world. Consequently, local weather bureaus did not have to wait a long time for the daily picture of the earth before drawing up their forecasts. The satellite's orbit for the following day was transmitted daily from New York. The APT stations received these data, stored them on punched tape, and then used this tape to automatically transmit the information to the receiving antennae. The Essa system was so ingeniously conceived that it worked satisfactorily for years. Naturally, one of its main drawbacks was that it was not possible to observe the earth at night.

Military Weather Satellites

The meteorologists wanted pictures of the earth that showed as large an area as possible, since they were not interested in details, but rather in overall weather conditions. A very high resolution power was therefore unnecessary. Their first step was to orbit the Essa satellites at an altitude of 875 miles, as compared to Tiros satellites at about 440 miles. It was sufficient if they could identify objects of a magnitude of two miles. Additional plans envisioned weather satellites at several thousand miles altitude, but these were never realized.

This development was understandable, but did not please the military, as they had profited greatly from the weather satellite pictures over the years. The American Air Force especially had based all its worldwide activity entirely on these pictures. Here we must also include their espionage satellites. From the photographs of the Tiros satellites they could tell which areas of the world were cloud-free, and could program the cameras of these spies in outer space accordingly.

The military also needed weather satellite pictures for other purposes. For example, the crew of a U.S. Air Force long-distance bomber is to rendez-vous with a tanker plane over a remote area of the Pacific. Because it increases their range enormously, midair refueling of bombers is one of the mainstays of the American Air Force.

The bomber is speeding towards its goal in bright sunlight, not suspecting anything. Then it suddenly gets the warning: turn around and proceed to a neighboring area. Tiros 1 had reported that the rendez-vous area lay on the edge of a typhoon, while showing at the same time a large fair weather area in the immediate vicinity. Since there are no weather observation stations in these parts of the Pacific, this warning was only possible by satellite.

For such undertakings it is sometimes essential to know the weather in a definite, sharply defined area. It is not surprising then that the military did not like the plans of the meteorologists, who were opposed to increasing the resolution strength. In the mid sixties, the military made themselves independent; they stopped using the Essa satellites, and constructed their own weather satellite system.

The Americans launched their first purely military weather satellite on January 19, 1965. Like its predecessors, it flew over the earth at an altitude between 440-500 miles, and had equipment to photograph the world day and night, in both optical and infrared wave lengths. For a long time the existence of these satellites was kept secret. It was not until March 1973 that John McLucas, Under Secretary of the Air Force, first officially revealed that the military owned a weather satellite for the specific purpose of supporting worldwide Air Force missions.

Most of the details of the camera technology are still secret. All we know is that the infrared cameras can distinguish objects with a diameter of two miles, while the optical cameras can make out objects with a diameter of only 1,500 feet. This must be extraordinary equipment: even the optical cameras are capable of photographing in pitch-black night! On the pictures we can see brightly lit cities and villages clearly. Indeed, their sensitivity is so great that bright fires on the earth can cause overexposure. Apparently the techniques used here were developed for the war in Vietnam. The illumination from stars is sufficient to provide clear pictures of the ground.

The pictures are received by ground stations whose locations are kept secret. In part, they are moveable stations which can be taken wherever needed, making the system extremely flexible. Within minutes those responsible get the information they need to make crucial decisions. At the same time, the pictures are relayed via the military communications satellite network to the U.S. Air Force's world weather center, where they are put in digital form. These signals are then fed into a computer which automatically evaluates the pictures. In this way, the subjective influence of human beings is completely eliminated.

The Soviets started using military satellites in 1962. This does not mean that they concentrated on the military uses of space exploration more intensively than the Americans; in fact, quite the opposite. For example, because they had difficulties in developing the technology, their satellite espionage program started much later. The Americans prepared their espionage missions by using Tiros satellite pictures to choose targets not hidden behind clouds, while the Soviets were faced with the problem that they did not possess any civilian weather satellites. Their camera technology was at this point far behind that of the Americans.

In order to prepare their espionage missions in a similar way, the Soviets had no alternative. They had to launch their weather satellites into orbits much lower than those of the Tiros satellites. In this way, at least they could compensate for the poorer resolution of their cameras. This naturally meant the satellites had a much shorter life expectancy, so meteorologists were not particularly interested in them. For this reason, the Soviet Union developed a military weather satellite system very early.

The Perfect System

It took quite a while before the Soviets had advanced far enough in their camera technology to turn to building their own meteorologically oriented satellites. The launch of Molniya 1 on April 23, 1965 marks the starting point. We have already met the Molniya series in their function as communications satellites. Actually they are multi-purpose, like NASA's ATS satellites. One of their tasks is to relay pictures of the earth from a great altitude which are then evaluated by meteorologists.

The quality of the pictures from Molniya 1 was worse than just bad, so they had to try something else. Using their Cosmos satellite, the Soviets developed a partial program dedicated to meteorology. It is difficult to say exactly when the first Cosmos weather satellite was launched. We know for sure, however, that Cosmos 122, launched June 25, 1966, belongs to this series. It orbited the earth at about 400 miles.

Cosmos 122 probably had predecessors, but the Soviets are still secretive about their activities. We presume that these earlier satellites were not equipped with television cameras, but instead took photographs on film which was then sent back to the earth. If this was the case, the pictures could never have been included in regular weather forecasts.

At any rate, Cosmos 122, at the very latest started the Soviet weather satellite program. In the same year that the satellite was launched, the Soviets signed an agreement with the Americans providing that information from weather satellites be regularly exchanged between Washington and Moscow. The two great power blocs of the world were thus able to overcome their political differences and cooperate, at least in the field of meteorology.

Just as in the United States, several years passed before Soviet weather satellite technology was far enough advanced to implement a routine operational system. This system was given the name Meteor, and its first satellite, Meteor 1, was launched on March 27, 1969. With it the Soviets had finally found an adequate tool for satellite meteorology. Basically the Meteor satellites are equipped with two television cameras and an infrared camera, so that they can watch the earth day and night.

In spite of the bilateral agreements with the United States calling for a regular exchange of pictures, the Soviets are not ready to let the world see what they are doing. In fact, they have even warned the director of Germany's Bochum Observatory, Heinz Kaminski, against using his equipment to receive their pictures, although he had previously enjoyed good relations with Soviet scientists. They also threatened to turn off the equipment in their Meteor satellites within the range of Bochum, if he did not heed their warning. They have not yet given an explanation for this strange behavior.

Nonetheless, the Soviets did remain cooperative in the area of satellite meteorology. Together with the Americans, they regularly provide Moscow, Melbourne and Washington, the world centers of weather investigation, with photographic material and other weather satellite data. Thus an old dream of the meteorologists, the World Weather Watch (WWW), has finally come true. Every scientist concerned with meteorology or climatology has access to the results of the worldwide weather satellite reconnaissance systems of the East and the West.

Meanwhile the Americans have further increased their contribution to the WWW, through their Weather Satellite

Operations System. The Essa satellites have been replaced by a new generation of satellites. Even at the time of the Tiros satellites, the Americans knew that in the long run an operations system would actually have to deliver much more information about weather than would be possible with the Essa satellites. These deficiencies could only be eliminated with a more advanced system.

The precursor of the new system was a satellite which had long been planned under the cover name Tiros M (The capital letter M means that the satellite had not yet been launched.) On January 23, 1970 the satellite was finally orbited at an altitude of 875 miles. Now it finally got a name: ITOS 1 (Improved Tiro Operational Satellite).

The ITOS satellites differ from the Essa satellites in primarily two ways – in their means of stabilization and in their hardware. Their stabilization is based on that of the Nimbus satellites, and their photographic equipment always points towards the earth. The Americans have made so much progress in technology in the past few years, that they could employ this system routinely.

In reference to the equipment of the ITOS satellites, we must mention that they combine the capabilities of the two Essa satellites – direct broadcast of photographic material plus storage on a magnetic tape. Furthermore, they have taken on a whole series of additional tasks. These satellites are particularly good at taking infrared pictures, which they can also relay to the APT stations.

The ITOS satellites also have equipment for making measurements of the earth in different areas of the spectrum. As we have stated, cloud observation alone does not allow for any concrete data concerning temperature and air pressure in the atmosphere, or concerning the other parameters which have such a great effect on the weather. Since the equipment has already been tested in other satellites, the time has finally arrived to expand this research by using satellites. We will discuss this in greater detail later.

ITOS 1, in spite of its somewhat confusing name, was a test satellite, and consequently belonged to NASA. It did not take long for the tests to be successfully concluded, so that NASA could hand over this new satellite to the U.S. Weather Bureau. On December 11, 1970 the next generation of American weather reconnaissance operations satellites was put into service with the launch of NOAA 1. This stands for the National Oceanic and Atmospheric Administration, the agency which includes Essa, but which in turn is under the Department of Commerce.

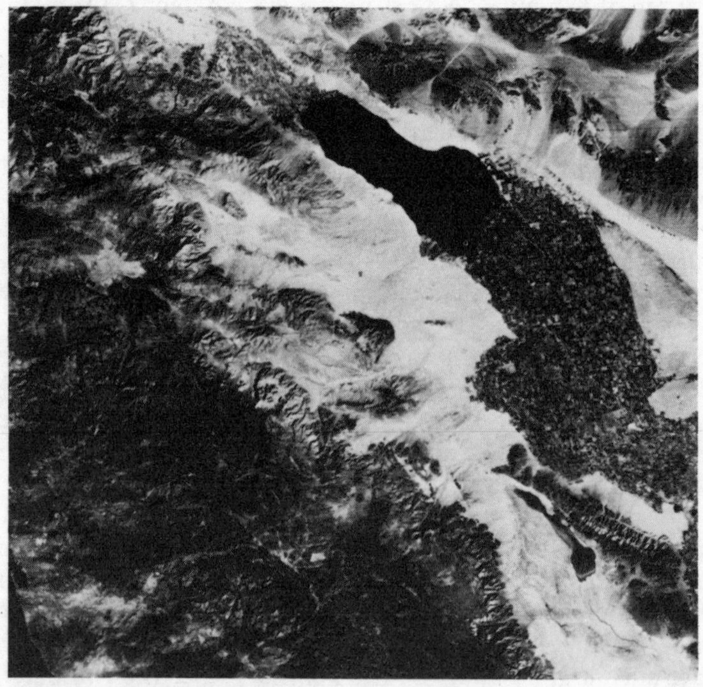

Testing grounds in California. Illustrative is the cultivation in the Imperial Valley south of the Salton Sea. The border with neighboring Mexico is a sharply defined straight line.

9. The Large-Scale Model

Measuring The Earth's Radiation

There are two safe ways to arrive at a weather prediction. One is to base the forecast for the next day's weather on the information gained from previous weather observations. This is the more primitive method. The number of weather observations which must be made has grown so rapidly that computer evaluation of the data becomes inevitable.

The second way is to construct a mathematical model for weather development. When events in the atmosphere can be represented by formulae, a large computer will be able to produce an automatic weather prediction. Of course, it must first be programmed with relevant initial data – temperature, air pressure, wind speed and so forth for various places on the earth at a predetermined time. With sufficient data and an exact knowledge of weather patterns, one could thus not only make a more accurate immediate forecast, but also a reasonable long-range forecast – about one week in advance. The advantage would undoubtedly be enormous. In recent years, important progress has been made in this direction. Weather satellites have improved our knowledge of the atmosphere to such an extent that the necessary mathematical model is no longer a dream. In 1974 meteorologists began the first large-scale experiment designed to prove that they are heading in the right direction. The art of weather forecasting should improve markedly. Even though the entire undertaking is still in the testing stages, and practical use can only be expected some time in the future, the development which has already started is so important that an explanation in broad outline seems in order.

As already mentioned, the sun plays an important role in weather development. It is the number one source of energy. Part of the solar radiation reaches the earth and part is swallowed up by the atmosphere. The portion which penetrates to the earth's surface warms the land and the oceans, and provides the energy for all forms of life, as well as many other processes. To a certain extent it is also reflected, and thus returns to outer space. In this way, even more of the radiation is caught in the atmosphere. It is important for meteorologists to know exactly how much energy is actually absorbed by the atmosphere during this entire process.

At the end of the fifties, American scientist Verner E. Suomi of the University of Wisconsin addressed himself to this question. Fortunately he was acquainted with space exploration technology which was then rapidly developing, and he found a simple way to arrive at his first tentative conclusions. He suggested mounting two hemispheres on the outside of the satellite, one black and the other white. Due to the different properties of the colors, the black hemisphere would absorb almost all of the radiation which struck it, while the white hemisphere would absorb only long-wave radiation. Since long-wave infrared radiation is nothing more than heat radiation, the earth's radiation balance could thus be roughly measured with these two hemispheres. Suomi's idea was proven correct in October 1959, when the "Suomi detector" affixed to the outside of Explorer 7 delivered satisfactory results.

The meteorologists were of course not satisfied with such a simple means of measurement. Since the principle was proven to be correct, it had to be developed still further. This meant creating new equipment to research the earth's radiation content in detail. Where exactly was the energy pumped into the atmosphere, and what effects did it have? Question after question arose, and we still have not answered all of them.

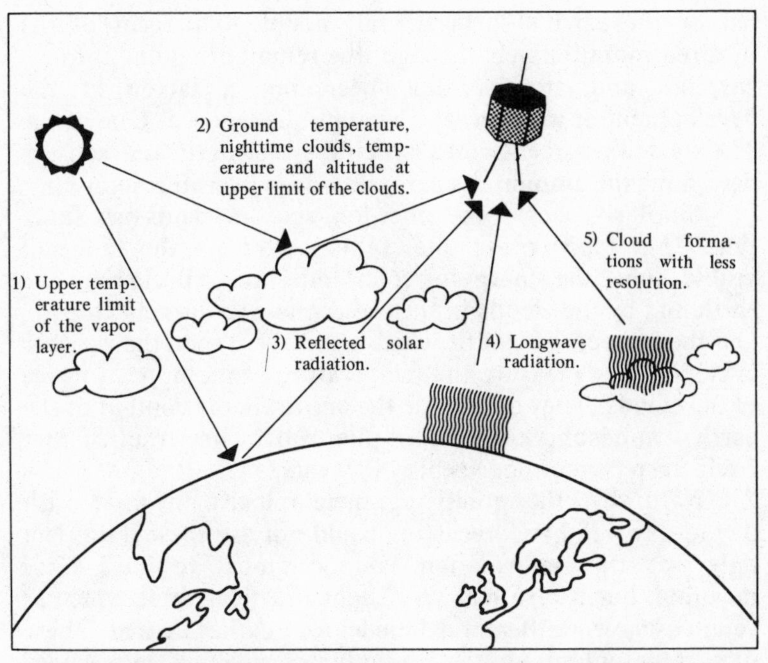

Infrared detectors, with which some of the Tiros satellites were equipped, provided valuable information about the earth's radiation content. They measured the innate radiation and reflected radiation of the ground and of the cloud surfaces.

The Tiros program did provide many of the answers. Some of the Tiros satellites (as their name, Television and Infrared Observation Satellites, indicates) were equipped with five infrared sensors which were to be used in meteorology and basic research. Each of the sensors measured the earth's radiation in a specific wavelength, so that when all the measurements were combined they provided a more exact picture of the earth's radiation.

As an example, we will describe how one of these five infrared sensors works. It is a well known fact that water vapor in the atmosphere swallows heat radiation of a defined wavelength. This applies especially to the heat radiation emitted by the earth's surface. Only about 40 percent of the infrared radiation gets through, the remainder going into air currents and storms, thus becoming a factor in the development of weather which cannot be ignored. One of the Tiros satellites reacts to a specific wavelength and sensors determine the amount of energy absorbed by water vapor.

Similarly, each of the other four sensors has its own function. They register the total radiation within the range of visible light, the energy of solar radiation reflected by the earth and by the clouds in the atmosphere (known as albedo), and the temperature of the earth's surface. When the weather is cloudy they measure the temperature of the upper surface of the clouds. They determine the entire energy content of the earth – atmosphere system. A pile of data has resulted that could keep meteorologists busy for years.

Naturally, the practicing meteorologist involved with day-to-day weather forecasting could not use these data. Not only was this information far too global to have local meaning, but it also had very slight practical value, since as yet no useable mathematical model for weather existed. These data represented only the beginning of satellite meteorology development. In another respect, however, they were of practical use. The meteorologists realized that, because the power of their detectors was very small, they saw the earth much too coarsely for many purposes. Thus, for the Nimbus and later satellite series, they developed infrared equipment with a strong power of resolution, capable of examining the earth line by line. This makes it possible to follow the earth's cloud covering even at night.

Knowledge of the earth's radiation alone, regardless of how valuable it might be, will not do much for weather prediction in the future. It is much more essential to understand the various parameters of our atmosphere in their totality. Using temperature as an example, one can understand how difficult this is. When there were no clouds in the skies, the Tiros satellites measured only the temperature of the earth's surface;

when the skies were cloudy, they measured the temperature at the top of the clouds. The measurements were thus more or less two-dimensional. The atmosphere itself, however, is three-dimensional: at height of six miles, for example, the temperature is different than it is at three miles. Similarly, humidity, air pressure, wind speed, and wind direction all vary with the altitude.

In the sixties meteorologists attacked the problem of testing the earth's atmosphere in its entirety. Balloons and high altitude rocket shots were used only sporadically, because they could not yet provide continuous surveillance. Would it be possible to make such a sounding by satellite, and if so, how? Or did they have to develop an enormous rocket and balloon program of gargantuan dimensions?

The longer one ponders this question, the more impossible it seems that the atmosphere can be tested by satellites, since after all they fly high above it. (What we mean here by atmosphere is the lower part of the atmosphere where weather takes place.) Yet testing is possible with the help of spectrometers. In 1959 the well known scientist Kaplan suggested building spectrometers into satellites.

In principle the way spectrometers work is quite simple. The various components of the atmosphere, especially water vapor, nitrogen and ozone, prevent radiation from the sun in certain wavelengths from reaching the earth's surface. They are absorbed by the atmosphere. Radiation in different wavelengths, the so-called "atmospheric windows", can penetrate the atmosphere unhindered.

Consequently, satellite measurements of radiation through these windows are focused on the earth's surface. However, where absorption takes place, they are focused on high altitude areas. By appropriate selection of the bands in which the measurements are taken, scientists are able to determine the atmospheric conditions at, for example, three, six and ten miles altitude, just by looking at the bands of absorption which occur at these altitudes. From these measurements the temperature at these altitudes can also be determined, among other things. The precise steps to be taken in this process are of no interest to us here.

Satellite Nimbus 3, launched in April 1969, took the first approximate space measurements. Subsequently, Nimbus 5 was equipped with a spectrometer capable of determining the temperature of the atmosphere at seven different altitudes. The measurements included areas from the earth's surface up to thirty miles. They provided the meteorologists with new, surprising knowledge which will surely be of future value to the science of weather prediction.

The entire lower atmosphere could now be tested daily – a new milestone. Since 1972, these measurements have been available to any interested party. Temperature maps of the earth have been published yearly with atmospheric temperatures at altitudes of thirty miles. At present, it is true, they do not have any great importance for the practicing meteorologist. The British Weather Service, for example, showed no interest in the maps.

This example of a weather satellite taking measurements in space shows clearly the stage reached by modern measurement technology. We have discussed this technology in greater detail, because atmospheric temperature is one of the factors about which we must have regular global data if we want to make weather forecasts based on a mathematical model. To this extent, these measurements differ from other purely scientific measurements whose practical application is difficult to foresee.

We must mention that these data are only a small part of the Nimbus satellite program. Nimbus 5 alone was to carry out six experiments, including the most varied measurements of cloud, water, and land temperatures. Among other things, the temperatures of the earth's crust were also determined. This was supposed to provide data of value not only to meteorology, but also to geography, oceanography, agriculture and other disciplines. Unfortunately, the equipment failed after only three and one half weeks. A device is now being built for Nimbus F which will test the atmosphere at twenty different levels, up to an altitude of fifty five miles.

Ghost

The advances made in recent years in the area of atmospheric testing are astonishing. Nonetheless, although satellites can provide much information necessary for predicting the weather, they cannot provide enough. We are still unable to utilize to the fullest mathematical models of weather development. At least one aspect keeps evading all attempts at observation from outer space: data concerned with wind speed and direction.

Successive photographs taken from satellites show the movements of clouds, but this is not necessarily identical with the movement of the winds. In order to obtain all weather parameters essential for an exact prediction, we must turn to other methods. The logical step is to go back to traditional means – balloons and rockets.

In a press conference of the International Transportation Exhibition in Munich in August 1965, Wernher von Braun first mentioned an enormous project which NASA had been considering, and in which all countries of the world were to participate. The entire earth, including the oceans, was to be covered by an evenly distributed network of high altitude meteorological research rockets. The cost of this system would be divided among the individual countries, and determined according to the size of each country. The rockets could then be launched at regular intervals, and all data concerning weather and atmospheric conditions important to forecasting would be measured with various instruments.

Actually the project was much more imposing, since they intended to automate as many of the separate processes as possible. For example, part of the observation system was a satellite designed to be a data collector as well as a command station. Every time one of the rocket ground stations entered its field of vision, it was supposed to send out a command which would automatically cause a rocket to be launched. At the same time that this rocket was climbing, the satellite received and gathered the measurements.

This was not all. The satellite was to perform yet another task. With a miniaturized computer, it was to immediately begin working with the figures to provide a weather forecast. This forecast would then be corrected and updated continually, using the most recent measurements, and the optimal goal would be attained: in every country on the earth within range of the satellites, the most accurate weather forecast would be available simply by pressing a button.

Theoreticians have backed such a system for a long time. Their calculations have shown that it would work, at least in principle. If the rocket stations were 300 miles apart, it would be possible to predict the course of the weather for two weeks in advance, using two atmospheric tests per station per day. During the summer months this forecast could be expanded to eighteen days.

In practice, however, there were numerous difficulties. Leaving out lesser problems computer technology is still not able to provide the small, powerful, competent electronic computer essential to the project. It is also questionable whether the individual countries are ready to bear the enormous cost of the system. Rocket launches are not cheap. Even if they were to become routine, they would probably still remain expensive enough to scare off many a politician.

Naturally, many wondered at first whether using balloons would not make the system cheaper. Balloons do not pass through the atmosphere in a few minutes, but can stay within it for hours or even days. At first glance, this might seem to be the answer, but conditions are not quite so simple. As soon as the temperatures in the atmosphere start to rise – and the temperature varies considerably during the course of a day – a normal balloon expands. This does not affect its weight, of course, but its density suddenly decreases. Consequently, it rises and finally bursts. Before this happens, it shoots up and down through the atmosphere, so that its measurements can never be related to a constant.

Nonetheless, with a simple adaptation, balloons can offer the key to success. Instead of a normal balloon, one could use a high pressure balloon whose volume is held constant by high interior pressure. The interior pressure of the balloon is thus always greater than the exterior atmospheric pressure. When the interior pressure equals approximately 25 percent of the exterior pressure during the day, there is still a smaller interior pressure during the night. This means that the balloon enevelope never gets limp — unless it gets a hole in it. Balloons can thereby float around in the atmosphere for weeks at a level where the atmospheric pressure is constant. This makes it possible to include all measurements in a mathematical model.

A worldwide system of such balloons would only be practical if there were a satellite to receive the measurements; data concerning wind speed and direction can be determined most easily by the drift of the balloons. This means their location must be measured very carefully with an error of only a few miles at most. Only satellites can attain such precision on a worldwide scale.

This fact led to project GARP (Global Atmospheric Research Program), in which freely floating balloons regularly test the atmosphere at various altitudes. They then transmit these data to a satellite which relays them to a ground station once each revolution. Evaluation of the data by the satellite itself is still in the making.

We cannot deny that such a project raises many extremely difficult problems. For example, a balloon designed to float in the atmosphere for weeks must be extraordinarily stable. Furthermore, it has to be lightweight, while at the same time it must be able to carry much data-gathering equipment. Technicians have successfully built large balloons which are several meters across and weigh no more than

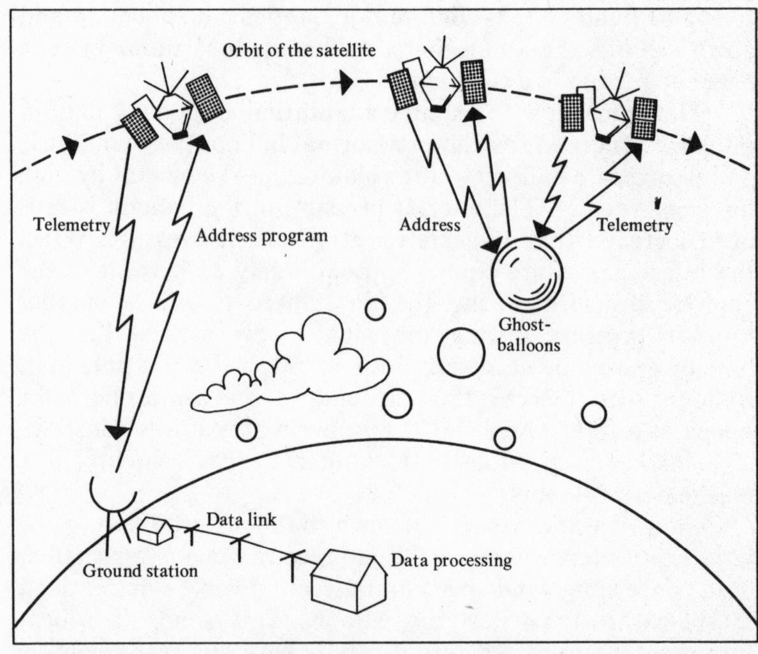

Project Ghost: a network of several thousand high pressure balloons test the earth's atmosphere. The measurements are transmitted to a satellite which relays them to a ground station. A geostationary satellite can be used for faster transmission of the data.

eleven pounds, in spite of their measuring instruments! This was made possible by using Mylar foil for the envelope, just as with Echo 1 and Echo 2.

Extensive data have already been obtained for evaluation by GARP. The idea of such a program is more than a decade old. It is the realization of two suggestions. Years ago, the French worked out a project under the name "Eole", which was limited to the southern hemisphere. A total of 512 high pressure balloons were supposed to test the atmosphere there, where there are few weather stations, and then transmit their findings to special data collecting satellites.

The planning for the American project "Ghost" (global horizontal sounding technique) of the National Center for Atmospheric Research (NCAR) at Boulder, Colorado is on a much larger scale. The balloons in the Ghost project will not be confined to the southern hemisphere, but are to fly over the whole earth. Six to ten thousand balloons are necessary to gather enough data on temperature, air pressure, and humidity at several different altitudes. Wind and speed direction can also be ascertained from the drift of the balloons. Here too a satellite will collect the data.

In March 1966, NCAR made its first test launchings of Ghost balloons. An area not far from the city of Christchurch, New Zealand was the launching place. In the ensuing years the balloons have continued to rise from there, and their number has now reached several hundred. Many of them, especially at the beginning, were written off as failures, but the number of successes has gradually increased.

Even in the first days of the experiment, a few successes were achieved. The scientists could barely believe the flight of the balloon launched on April 10, 1966. Slowly it rose from the launching platform and climbed higher and higher, until the ball with a diameter of 85 inches finally reached its intended altitude of 7.5 miles. Then its sensational flight began. Without pause, the wind drove it towards the east. The balloon flew over the South Pacific and South America, crossed the South Atlantic and Australia, and, after ten days, reached the Fiji Islands. For the first time, a balloon propelled only by the wind had flown over the entire Antarctic. However, that was not the end: it circled this continent of ice and snow six more times, until radio contact was lost seventy four days later.

Another Ghost balloon relayed its readings back to ground stations for over a year, a record which has probably not been broken. It was also launched from Christchurch, and was blown by the wind for 467 days until its radio equipment ceased functioning. With these successes the Global

Atmospheric Investigation Program, using high pressure balloons, had advanced an important step. What was mainly lacking in the test stages of the experiment was the use of satellites as data collectors.

Satellites as Data Collectors

Collecting data by satellites, as it is presently being tested, is primarily based on an American system called IRL-"Interrogation, Recording and Location". Scientists had originally planned to begin using it in 1968. The experiment had to be postponed for several months, however, because of the abortive launch of the satellite Nimbus C. Zoologists were the first to use this new technique. As a matter of fact, they had already made known their desire to use satellites as data collectors long before this. Ethologists in particular had a series of questions which they either could not solve, or could only solve with great difficulty using traditional methods. For example, it was easy to follow the short migrations of large animals when they were equipped with small radio transmitters. Because these transmitters only had a range of about sixty miles, it proved to be enormously difficult and troublesome to track the animals at greater distances.

Furthermore, there was no way of tracking animals in the more isolated areas of the world, for instance in the polar regions.

At first glance such questions seem to have nothing to do with space exploration. However, it was thought that by observing animals during their long migrations they might reveal their methods of orientation, or, as with whales, their methods of navigation. If this could be mechanically simulated, it might lead to new navigational systems. This is one of the reasons why NASA was ready to cooperate with the zoologists.

Looking into the future, regardless of how far away it may seem, NASA also had a certain interest in recording the hibernation habits of several animals, as well as their physiological behavior during hibernation. What is the secret of hibernation? Will it perhaps one day be possible to make people fall asleep artificially for a certain amount of time? Or will this remain a dream forever?

In 1972 NASA workers and zoologists together went to hunt animals, not to kill but for experiments. Among their quarry were elephants in Kenya, and black bears and elk in North America. As soon as they were caught, the animals were drugged and equipped with a small radio transmitter. Then the actual experiment began: broadcasting the data to the Nimbus satellites.

For thirty days Nimbus 4 registered the animals heartbeats and body temperatures. The black bears were hibernating at the time, while the elk continued to cross the land under constant satellite observation. Their position was determined within a range of about five miles. Thus the problem of collecting data with the help of satellites was solved. (Incidentally, the equipment of the Nimbus satellites was specially designed to gather data from more than 100 transmitters on earth. It made no difference whether they were on animals, in weather balloons, or built into buoys.)

GARP was planned to be tested on a large scale in 1974. Two hundred planned measuring stations are to be scattered over the entire Atlantic, and will send their readings to a Nimbus satellite. Worldwide application of this system will then follow in 1977. According to present plans, the final version of the data collecting satellite circling the earth will obtain its data not only from high pressure balloons, but also from buoys, ships, airplanes and ground stations. Perhaps it will not even have to send its data back to a ground station, but will be able to use one of the geostationary satellites. In this way, all measurements will reach the ground stations as quickly as possible: sent via geostationary satellites they will not have to be stored first. Nimbus F, also planned for launching in 1974, is to be equipped with an antenna for communication with the ATS satellites.

Much more work must be done before GARP can finally

be put into operation. The system has raised many problems. For example, when passing over the earth, the satellite automatically starts the individual transmitters at the measuring station. In order to do this, it has to know the approximate position of the balloons in advance, since it quickly passes out of their range. What will happen if one of the positions is incorrectly calculated because of a faulty estimate of the wind? How will the satellite find the balloons again? Until the meteorologists have solved all of these problems, they will not know whether their dream can be realized: namely, long-range weather forecast with the help of a mathematical model.

Geostationary Weather Satellites

Even when GARP is finally in use, local forecasting cannot omit weather satellite observations of the earth's cloud covering. For this reason, development of weather satellites has not ended. Despite the superb functioning of the NOAA satellite system, it does not fulfill all the desires of meteorologists.

One deficiency in particular annoys the weather prophets. Short term changes in the weather, for example, sudden developments in hurricane systems, can only be recorded incompletely by the operations satellites. With two NOAA satellites flying around the earth simultaneously, a picture of a specific area can only be made every six hours at the most. What happens in the meantime is hidden from the satellites.

This is nothing new. Scientists realized this by the mid sixties, and developed a new technique which was soon to reach perfection: the use of geostationary weather satellites. These satellites have the advantage that they can watch an enormous area almost continually, since they are quasi-stationary over a specific point. NASA's technological experimental satellites (ATS) have already shown the value of this round-the-clock supervision. One of them was even moved over the Caribbean for a short time during the hurricane season to keep constant watch on the hurricanes in the Atlantic.

ATS 1 was the first ATS satellite equipped with television cameras for observing the earth. It was stationed over the Pacific, where, from its elevated position, it took a picture of the entire Pacific every twenty three minutes. It provided the scientists with research material that they had previously only dreamed of. From its pictures they could easily put together a plastic film showing the dynamics of weather development over a third of the earth at a single glance.

Anyone who has seen one of these films will not soon forget it. Most impressive is the play of hurricanes – their movement and their growth. These huge, energy-rich monsters powerfully contract their spirals and suck everything they can reach into their maw.

Meteorologists were especially surprised that they had previously held a completely false picture of worldwide cloud dynamics. They had assumed that the equator was a barrier for the large cloud systems, and that the weather in the northern hemisphere was almost completely independent of the weather in the southern hemisphere. As it turned out, this was not the case. Even spirals of monstrous proportions crossed the equator. There was no natural boundary. This idea was simply the product of fantasy, and the desire to show the development of weather in a simplified manner. We can work with a model only if its parameters are known, and the equator would have provided an elegant boundary.

Weather conditions in equatorial regions, despite their striking visibility, do not have as great an influence on the weather in the temperate zones as had been thought. Although the equator is not a meteorological wall, meteorologists can in practice almost forget the equatorial region in their investigations. In the final analysis, we must introduce some simplifications in order to understand weather at all, and any other simplifications would have a far greater effect on the experiments.

In any case, it has become clear that geostationary weather satellites have advantages which the NOAA satellites do not. This is true not only for practical weather forecasting, but also for purely scientific meteorology. The Americans have therefore joined with the Europeans, the Japanese, and the Australians to expand their operations satellite program. In a few years, they will put into use a system of operational geostationary weather satellites. The United States will provide two satellites, Europe a third, and Japan and Australia together a fourth. The result should be another large increase in the material that meteorologists receive for their weather forecasts, and consequently, a further increase in accuracy. This in turn will help prevent much damage and loss, and will thereby benefit the individual citizen.

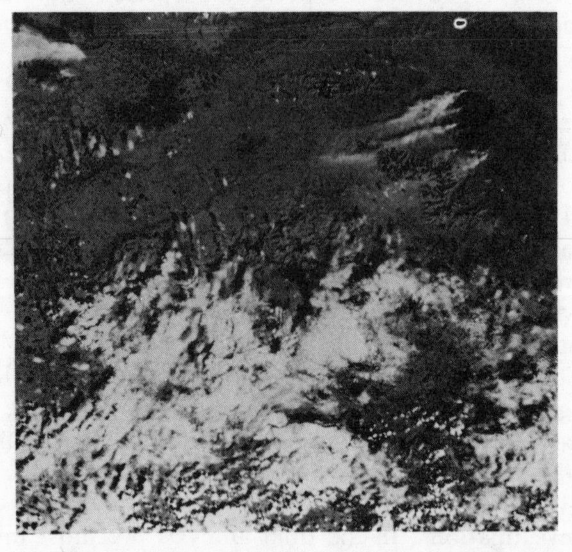

A forest fire in Alaska. The smoke carries across the land for miles.

10. The Way into the Future

At the Limits of Perception

"In the highlands around the Himalayas, I saw individual houses with smoke coming out of their chimneys. On the border between Taxas and Arizona I saw something on the street that looked like a truck. Once I even saw a steam locomotive in northern India. First I saw the steam, and then the train. Near Burma, I could pick out the wake of a ship on a broad river."

When American astronaut Gordon Cooper described his Mercury flight around the world in May, 1963, he ran into a wall of disbelief. Was he trying to be a modern Munchhausen, or had he just fallen victim to hallucination? The scientists who listened to him at that time thought the latter. In any case, they could not believe his stories. They argued that noone could recognize such fine details with the naked eye from the altitude at which the space ship was circling the earth, not even Cooper, although tests showed that he possessed unusually sharp eyesight.

Yet the scientists gave Cooper a chance. They asked him to say exactly when and where he had made his sightings. Thanks to his good memory, he was able to do this precisely. Once again, the scientists were astonished: they found that all his stories were true! For example, at exactly the time specified, at precisely the place on the border between Texas and Arizona where Cooper claimed to have seen something like a vehicle, a glaring white truck of the Border Control had stopped. Cooper's credibility was restored.

NASA then carried out a number of experiments within the manned Gemini program designed to help explain the phenomenal visibility of small objects on the earth from orbit. During the flights of Gemini 5 and Gemini 7, test discs of various sizes and markings were laid out north of Laredo and near Carnarvon, Australia. The astronauts were to look for these. Rockets and airplanes were also sent aloft, especially for the spacemen.

Radio conversations between the astronauts and the ground stations indicate that some of the tests, although not all, must have been successful. During the flight of Gemini 5, Charles Conrad said from his capsule: "It's wonderfully clear today. . . I see all the streets in Jacksonville, Cape Kennedy, and on down to Miami. I have also spotted three airplanes taking off from Jacksonville." During one of the next orbits, shortly after the launch of a Minuteman rocket from Vandenburg, California, he called quite excitedly: "I see it, I see it. . . There goes the rocket. . . over the water."

Unfortunately the results of these naked eye observations were for the most part kept under strictest secrecy. Some of the photographs the astronauts took with wide-angle lenses showed various airports and their landingstrips so sharply, that they were artificially distorted before being published. The military, which took part in these experiments, did not want anyone to look at their secrets.

Secrecy is the catchword. At the beginning of the sixties, the military in particular were interested in detailed photographs of the earth taken from orbital altitudes. They accelerated this development with all available means, and this led to an almost continual supervision of our planet. Techniques which would have seemed impossible to civilians were the final results. It is now possible to photograph the structure of mineral deposits in our world. This will undoubtedly contribute to solving such pressing problems as providing food for our children and grandchildren. It will also help in the search for new sources of raw materials.

Perhaps things would have turned out differently if in 1955 Soviet Minister Bulganin had not turned down President Eisenhower's suggestion to allow every country to carry on as much air reconnaissance over any other country as it wished. This idea of sanctioned espionage could only occur to Americans, who in any case were now forced to develop espionage satellites for the task! Without a thorough knowledge of the state of Soviet rocket technology, atomic technology, etc., the Americans felt ill at ease. Many questions could only be answered by spying on the Soviet Union from above.

On January 3, 1961, after preliminary tests in the Discoverer satellite program, the first American espionage satellite, Samos 1, attained orbit around the earth. Samos is an abbreviation for Satellites and Missiles Observation System. Samos 1 was the first of a number of espionage satellites which have been launched more or less regularly since 1961. (The first Soviet espionage satellite was launched on April 29, 1962 from Tyuratam near the Aral Sea.)

Espionage satellites are an absolute necessity in the Cold War era. With only a small factor of error they can show the enemy's entire weaponry. This can help prevent nuclear war, which a country might possibly start out of ignorance of its enemy's strength. For this reason, in May 1967, President Johnson said unofficially at a meeting in Nashville, Tennessee that the military's space exploration alone had justified at least ten times the expense of the entire space exploration program. The cost at that time had already reached 35 to 40 billion dollars!

At this point, we should say a few words about military satellite espionage, since most of the television cameras and sensors which will lead to a revolution in space exploration are of military origin. Espionage in the optical wavelengths has doubtless had fantastic results. We can draw this conclusion from President Johnson's statement in Nashville: "Thanks to satellites, I know how many rockets the enemy has." As Secretary of Defense Thomas S. Gates later said, only "an idea of the transport problems and current location of armed units" was gained from the flights of the U2 espionage planes. These U2 flights over the Soviet Union are still shrouded in mystery. Satellites first made possible an exact census of the rockets in the Soviet Union and other parts of the world.

Even though the activity of espionage satellites is kept fairly secret, nonetheless some things do reach the public. The satellites themselves cannot be hidden, and with a certain amount of skill one can get an idea of what they are designed to do from their orbit in the heavens. For each satellite mission, whether military or civilian, a precisely determined orbit around the earth is plotted. In order to find out more about their activity, it is often useful to correlate the number of espionage satellites launched with various worldwide political events.

For example, during 1966 an extraordinarily large number of a certain American satellite were launched. This satellite apparently served for espionage, although this presumption was never officially confirmed. It was in precisely this year that two things happened in the Soviet Union to cast doubt on Soviet parity with the Americans in weaponry: the Soviets stepped up their development of large intercontinental missiles, and also began to secure Moscow and Leningrad against missile attacks from the West with an anti-missile belt. It is not difficult to guess the task of the American espionage satellites.

In a similar case, the Americans even admitted that they were using espionage satellites. In 1971, mysterious round holes were being dug all over the Soviet Union, especially in Siberia. At first their purpose was unclear. Finally, thanks to satellites, the mystery was solved. After a break in missile buildup, the Soviets were erecting new subterranean siloes for their so-called "armor piercers", intercontinental missiles of the S9 type. The destructive power of their missiles is twenty times that of the American Minuteman rocket. Without the help of espionage satellites this project would have gone undiscovered.

The Soviet espionage satellites also do not go unobserved. Nikita Khrushchev himself unofficially admitted their existence. He mentioned to Senator William Benton that he could place at his disposal pictures of American military support bases, and even once jokingly suggested exchanging them for American pictures. The main thrust of Soviet espionage is directed primarily against their Chinese neighbors, since the Soviets can keep good track of the United States by regularly studying American technical literature. It was no joke when an American once said, that with a certain amount of skill, a person could buy all the parts for an intercontinental ballistic missile on the free market.

We can see how seriously the Soviets take their Chinese neighbors by the fact that several espionage satellites put in orbit during the Ussuri River border disputes in 1969, at a time when they would normally have launched only two satellites per month. Shortly before the first skirmish the number suddenly jumped to two launches per week: the Soviets were expecting incidents along the border.

Many guesses have been made as to how much detail espionage satellites can really see on the earth. Today we can fairly safely state that the most recent American satellite type, "Big Bird", a collossus weighing eleven tons, is able to see objects with a diameter of only one foot from an altitude of about 150 miles.

Big Bird, or Lasp (Low Altitude Surveillance Platform) as it is officially called, uses various camera systems. With one of these systems it photographs large areas of the earth all at once. The pictures are radioed back to the earth, where they are immediately analyzed. If there is anything interesting to be seen on them, the satellite receives a command during one of its next orbits to photograph the corresponding area more closely with its telephoto lense. If necessary, an additional command can steer Big Bird into a more advantageous orbit.

Generally speaking, much is done by remote command. Every time a satellite approaches earth too closely because of the braking action of the outer atmosphere, its course correction rockets are fired. These push it back into a higher orbit, so that it can continue to work for two to three months unimpaired. Now and then it even expels film capsules which contain the highly sensitive exposed film. Before these capsules fall all the way to earth, airplanes with long nets catch them over the Pacific. If this maneuver is unsuccessful, a device ensures that these capsules disappear into the depths of the ocean without a trace. (The Americans probably remember with fright their experimental satellite Discoverer 2 which crashed near Spitzbergen while the recovery airplanes circled over the Pacific.)

Unfortunately, the Americans still refuse to release photographs from their espionage satellites. Indicative of this attitude is an incident which occurred soon after launching the space station, Skylab 1. Very serious difficulties arose, placing the success of the project in doubt. Among other things, the solar cell wings did not open. In order to examine the extent of the failure and to figure out possible ways to salvage the venture, Skylab was photographed from a U.S. Air Force observation station in New Mexico. The pictures were placed at NASA's disposal, with strict orders not to publish them. The reason given was that no unauthorized person was to know the photographic quality of the newest military equipment.

The Secret of Infrared Film

Sir Isaac Newton already knew that every action has an equal and opposite reaction. No country on earth will let itself be spied upon at will. With the help of suitable camouflage, it is possible to decrease the effectiveness of espionage from the air or outer space. We do not know how much use the great powers make of such methods, but they do not dispense with them entirely.

People say that the Soviets invented the helicopter as a means of troop transport. Helicopters do not leave as many traces in the air as do heavy transport vehicles, whose paths can be traced very precisely. There are even experts who claim that the Chinese built a whole series of fake installations in addition to their proving grounds at Lop Nor, where they test their atomic weapons. They did this to mislead the Americans and Soviets as to their true atomic potential. The green camouflage used in thick forests to make military equipment almost invisible from the air is almost primitive by comparison.

Today even these camouflage methods are not necessarily sufficient. A development of enormous consequences began during the Second World War, when resourceful scientists discovered that red was not exactly red, and that green was not green. If we photograph two red objects in a narrowly restricted color range by putting a filter in front of the lens, we can distinguish the two colors very clearly in the pictures. Although the eye itself does not notice any difference, almost no color is pure. The one color red might show a slight yellow tinge, the other a blue tinge, and these differences are clearly visible.

The scientists asked themselves why they should not be able to get more information about an object if they photographed it with a film sensitive not only to visible light, but also to the bordering areas of the infrared wavelength. (We must mention that this "near" or "optical" infrared has nothing to do with "far" or "thermal" infrared. The latter is responsible for the so-called heat pictures, and is used by weather satellites for night photographs. Infrared films can only be used during the day, since the near or optical infrared, just like visible light, is nothing more than sunshine reflected off objects in our world.)

One of the first scientists to work with this type of film was Robert N. Colwell, Professor of Forestry at the University of California at Berkeley. During the forties he investigated a fungus which caused great damage every year, attacking wheat and oats and destroying entire crops: the black stem blight. He was probably surprised himself when he looked at the developed film and discovered that the diseased grain

Spectrum	Ultra-violet	Visible light	Near	Infrared Mid and far windows	Microwave K, X, C, S, L, UHF, P
Wave length	> 0,29µm 0,4	0,7	0,9	3 5,5 8 14µm	1mm 0,75cm 2,4 3,75 7,5 15 30 100 136cm > 2m

The possibilities for remote observation could be expanded considerably over the course of the years by utilizing new spectral ranges. Besides visible light, the near and far infrared ranges, particularly, as well as the microwave bands (including radar), play an important role in space exploration.

could be easily differentiated from the healthy by its markedly darker coloration.

Basically, Colwell was ahead of his time. He gave the farmers a chance to discover the onset of the disease in entire harvests, before the human eye could see it, so that they could save the crops by taking appropriate early measures. In the case of the black stem blight, a dusting with sulphur will hold off the effects of the disease long enough for the grain to ripen. Most people, however, ignored Colwell's discovery.

Only the military took note. If it was possible to recognize black stem blight on infrared film, then it must also be possible to identify other plant diseases. Furthermore, it must be possible, above all, to differentiate organic from inorganic matter. This is indeed the case. It is precisely in the infrared wavelength that every plant has a characteristic appearance which is determined by its general condition. Modern color infrared film, in which the plant world appears in all shades of red, gives the pictures such an unnatural appearance that it is often called false color film. Dead plants appear black, and inorganic material appears blue.

The military saw that such film could make it possible to discover camouflage. Green camouflage nets suddenly become blue and stand out sharply from the red plantworld! Infrared film looks into the world of the hidden; espionage from the air and outer space has become much more effective because of infrared film. This was particularly true in

Vietnam where infrared film was used to find hidden paths and hiding places in the thick jungles. These could be discovered because of the dead and stripped underbrush.

Yet the capabilities of infrared film are limited. The main drawback is that it can only be used during the day. The military goal, however, could only be of round-the-clock reconnaissance of the earth. This naturally also meant night observation. Furthermore, additional means had to be found to obtain clear pictures of the earth from a great altitude when the skies were covered with clouds. Thus it is not surprising that over the years, specialists developed an arsenal of highly sensitive observation equipment to function in various ranges of the electromagnetic spectrum. They devised not only line-scanning machines which utilize thermal infrared to capture the smallest amount of radiated warmth, but also radar machines which can penetrate even thick cloud layers.

Radar reconnaissance is among the most refined methods of modern espionage. A radar beam is directed over the area to the side of the flight path of the plane or satellite. Reflected signals are received, and a clear picture of the area by day or night, in any weather, is obtained with the help of a complicated process of signal analysis. This technology is open to further improvement, but we can already foresee that in the next few years the advances made will permit recognition of even the smallest details from outer space.

The military were doubtless the pathfinders in the area of remote reconnaissance. It was inevitable, however, that the civilians would follow. Civilian scientists are still far behind in their development, but, they recognized at the latest in the early sixties, the possibilities which remote reconnaissance had to offer, and they then began intensive research. Their work culminated in a program which is bound to show the value of space exploration for global surveillance of the earth and all its natural resources. We must realize that the earth's treasures include not only gold, but also water, arable soil, fish in the sea, and many other things.

The purpose of EROS (Earth Resources Observation System) is not just a further development of purely scientific data. EROS is a test program specifically aimed at practical application. It must remain a test program until the best technology for observing the earth has been found. Even at the very beginning of satellite meteorology there were test satellites, the first Tiros satellites. The fact that these satellites relayed pictures to the earth which were immediately useful and valuable in practice (for example, the various hurricane pictures used in storm warnings) was incidental. This is also true with regard to the pictures from the first earth observation satellite.

Within a few years, the earth observation system EROS will undoubtedly evolve into an operational earth observation satellite system showing strong similarities to the present weather satellite system. Once again, as is the case with the practical application of exisiting satellites, this will show that Nobel Prize winners can be mistaken. It will prove that Max Born was wrong when, a few years ago, he disputed the value of space exploration by saying: "I do not see that space exploration will contribute anything to the material well-being of man." William A. Fisher, manager of the project, once explained the purpose of EROS with the following words: "We do not want geology, we want minerals; we do not want hydrology, we want water."

Nature's Fingerprints

In April 1966 William A. Fisher carried on an extensive conversation with reporters from the *New York Times* who were preparing an article on the employment of remote observation systems. He mentioned a picture the Gemini 4 astronauts had taken some time earlier. This picture plainly showed a large area of a certain developing country. It was so fantastic that the Americans thought many new sources of raw materials could be discovered in the country with its help, and they offered financial assistance for the project.

During the course of this conversation, Fisher mentioned something so important that we will quote it here: "Please don't publish the name of the country. The negotiations are very delicate, because this is the first time that the area has been photographed from the air." This showed clearly what detailed photography from outer space could offer. At the same time, the value of the work of scientists who had been taking "nature's fingerprints" was also confirmed. These men were helping to open the way to a new future.

Fingerprints are significant, unmistakable characteristics which serve for identification. If we want to photograph the earth from a spaceship so that the results are useful, then we have to know exactly how various phenomena – water, cultivated land, deserts, and so forth – will appear on the pictures. Amazingly enough, this is not at all simple.

For example, McDivitt, the commander of Gemini 4, had flown over El Paso, Texas in airplanes many hundreds of times. During his space flight, he suddenly found it difficult to locate El Paso. He kept looking for a narrow blue line, the Rio Grande. However, everything looked different from outer space. What he saw was a wide, elongated checkerboard pattern, and it took him quite a while before he realized that these were the fields along the river.

The Gemini 4 astronauts were also puzzled by a large triangle standing on its apex in Egypt. At first they thought it was a lava field of unusual proportions. Suddenly they realized what it was: not a lava field, but merely the fertile Nile delta. It stood out markedly from the dry desert regions surrounding it.

If the interpretation of ordinary photographs had caused problems at the beginning, how much more difficult would it be to interpret infrared photographs? Above all, the scientists wanted to get even more accurate information from these pictures from outer space. They wanted to be able to differentiate deciduous from coniferous forests, wheat from rye fields, and clean from polluted water. This was only possible if they knew exactly how the individual phenomena would appear on the pictures.

A program to produce a "fingerprint file" was already in progress at the time of Gemini 4's flight. The same problems which photography from outer space poses, had also to be solved for photography from airplanes. Since the fifties, airplanes had been regularly flying over certain areas. In the following years, they had started using infrared film on these flights along with regular film, since the former had proved very valuable. To interpret these pictures, however, they needed "fingerprints".

During the sixties a revolution quietly took place, equal to or perhaps even greater in importance than the introduction of fingerprints for identifying criminals. As an example, here are some figures: one day toward the end of the last century, the identity of twenty-seven prisoners was to be checked using photo albums. After fifty-seven hours, twenty-one employees of Scotland Yard had found only seven prisoners in these albums! When fingerprinting was later introduced, this same process was accomplished at the flick of a wrist.

The fingerprinting of nature, which goes back to Colwell, will have a similar effect. Colwell had recognized that on infrared film sick plants look different than healthy plants. If we develop this idea further, we must conclude that one should identify the reflective characteristics of the various wavelengths of plants, soils, water, and other objects in nature. Since these characteristics are different for each, thing in nature, we then have a basis of identification.

Thus the scientists in the sixties faced a twofold task. First they had to take an enormous number of fingerprints, and secondly, they had to find out in which wavelengths these fingerprints differed from each other the most. They intended to photograph the earth not only with color infrared film, but also in various narrow and separate wavelength ranges. The pictures in the other wavelengths provided additional information. They could easily be laid on top of each other to produce false color pictures, the so-called multispectral pictures.

In the first phases of the experiment, the scientists dragged heaps of sand, rocks and plants of every kind into their laboratories and measured their reflective properties. Mountains of fingerprints piled up in their files; they grew and grew. The scientists then swarmed out with their equipment, and began painstaking fieldwork. The fingerprints were not invariable. For example, how a plant reflects sunlight depends to a great extent on the surroundings in which it grows. Healthy plants act differently than sick ones; plants growing in soil rich in minerals act differently than those in poorer soil.

Finally they had gathered enough data to begin the actual experiments. Could these fingerprints they had recorded be recognized from great altitudes? They could not presume that this would be the case. Between the objects to be checked and the airplanes containing the measuring instruments, there was a thick layer of the earth's atmosphere. This has such a strong effect on the measurements that the fingerprints could indeed have proven unusable. Yet, after the first experiments, scientists breathed easier: the results were satisfactory.

At the same time, experiments carried on during the manned space flights showed that extending the tests to outer space was feasible. Photographs from there had such high resolution that large areas could easily be surveyed. Their borders could be clearly recognized, and often individual streets and bridges could be identified.

During the flight of Apollo 9 in March 1969, they finally tested the possibilities on a larger scale. They wanted to see whether they could get better information about the earth's surface by using satellites to take photographs within narrow spectral ranges. They used four Hasselblad cameras, each of which had a different filter. (Three filters admitted radiation from the optical part of the electromagnetic spectrum, while the other filter admitted radiation in the infrared part.) Using infrared film, they photographed predetermined areas on the earth. This was another complete success, and man had again advanced another step.

The following example will show how valuable these pictures were. Based on knowledge won over many years of research, all the data pertaining to the fingerprinting of vegetation were extracted from one photograph of an area near Phoenix, Arizona. All other elements in the picture were eliminated. What remained was a pure vegetation map of a large area, obtained entirely by automatic means. How much effort would it have taken to produce such a vegetation map by traditional methods?

This so-called thematic mapping of our earth holds great promise. Today it is possible to automatically convert satellite pictures not only into vegetation maps, but also into maps showing open water surfaces, snow or ice, or human settlements. The equipment for this is available to any scientist, and is located in the EROS center in Sioux Falls, South Dakota.

At the same time, surveys of the earth from the air were going on in high gear. Planes flew over broad areas, especially in the United States, more and more frequently. The air pictures were compared with ground measurements made at the same time. By the end of the sixties, scientists had succeeded in identifying plants of various kinds from the air with 95 percent accuracy, but they were not satisfied. Not only did they want to distinguish the different kinds of plants, but they also wanted to distinguish sick from health plants.

An experiment which NASA undertook in 1971 in cooperation with the Department of Agriculture, Purdue University, and the University of Michigan is worth mentioning. Fully equipped planes flew over seven states: Illinois, Indiana, Iowa, Minnesota, Missouri, Nebraska and Ohio. Two thousand corn fields in 210 test areas were photographed with normal and infrared film. The analysis of the pictures was supposed to show whether the course of corn blight could be followed by remote observation. Corn blight is a plant disease which mainly attacks cornfields, and it is precisely in these seven states that much corn is planted.

Naturally, in order to analyze the pictures they needed accurate information on the actual course of the disease. For this reason, 8,200 interviews were carried out from April to July 1971. Every farmer reported the condition of his fields, and the scientists then went to work. They needed twenty three days to photograph an area, and then to distribute their evaluated results to the interested parties. For the first time this disease could be observed over an extended area; from the photographs they identified three different stages of the disease! Scientists were now ready to test the use of satellites in surveying larger fields.

EROS

It may be of interest to discuss EROS in greater detail. We can say that it will provide valuable contributions to solving many of mankind's problems. Without EROS, these problems would have probably remained unsolved for many years to come. Basically we are already in the middle of our discussion: all the air reconnaissance flights carried out at the end of the sixties were not just a step leading to EROS, but rather a very important part of it.

A reason for further discussion of this subject is that earth observation from high altitudes is still in its beginning stages. While the first airplanes with infrared film and other reconnaissance equipment flew over various areas, precise field work was also being carried out. In this way, the scientists got an exact picture of how accurate the results of this air reconnaissance would be. Likewise, it would have been nonsensical to launch a satellite capable of photographing the earth with similar equipment without having carried on relevant comparative studies at the same time.

For this reason, EROS consists of three programs whose

value lies in being carried out together. Two of these programs (ERTS and EREP) deal with photographing from outer space, while the third (ERAP – Earth Resources Aircraft Program) is to supply comparison photographs taken by airplanes as they fly over the same test areas photographed by ERTS and EREP. This will enable the scientists to correctly evaluate the possibilities inherent in the two programs, without neglecting the field work.

Two technological earth observations satellites named ERTS are to be launched under the ERTS program (Earth Resources Technology Satellite). From an altitude of about 560 miles they will make detailed pictures of the earth in different wavelengths, and then send them back to the earth. Because they are automatically transmitted, the resolution power of the pictures is naturally limited, although the advanced technology of the equipment makes such extraordinary pictures that anyone would think they were normal sharp photographs.

The pictures of EREP (Earth Resources Experimental Package) taken from the manned space station Skylab provide a comparison in quality. Skylab is the same venture that made continual headlines because of various technical difficulties that occurred during its flight. When astronauts take the pictures, it is possible to work with film which can then be brought back to earth. Furthermore, since Skylab was only half as far from the earth as the ERTS satellites, it is understandable that the EREP pictures show much more detail.

It is still not clear whether this will really be an advantage in the long run; in the not too distant future the earth observations system will be developed to such an extent that we can begin thinking about its regular operational use. We will then have to find a way to digest the data which will be continually coming in. When ERTS 1 was launched on July 23, 1972, the scientists were optimistic. However, over the next few weeks it turned out that merely to send the pictures to all the interested people in the world created problems. Within one year the EROS data center in Sioux Falls, South Dakota sent out close to 800,000 ERTS pictures! No wonder that many had to wait up to a month to receive them.

We must not forget that each picture consists of 16–18 million information units. With a flood of forty pictures sent back to earth every day, this means that about 700 million information units reach the ground station daily. Furthermore, there are scientists who are still not satisfied with the resolution power, which incidentally is about 210 feet. This figure was obtained by studying pictures of the piers in Boston Harbor, where the exact lengths are known. Perhaps we should mention that limiting the information content of the ERTS pictures did not originally come about because of practical considerations in evaluation. Rather, the satellite builders were ordered by the American military to use observation equipment that would not distinguish any objects with a diameter of less than 35 yards. This restriction has now been lifted.

The problem of an excessive flow of data is not new. Years ago Samuel Koslov from the intelligence section of the Department of Defense pointed this out when he said: "Even if something is nice to own, do we really need it? You have to divide the cost of a system by the amount of useful information it provides, and you have to take into account how much of this we, the users, have to evaluate. What good are a million pictures when there are only ten people to evaluate them?"

Eventually there will be only one way out of this dilemma: automatic evaluation of the pictures. Machinemade thematic maps of the earth, a possibility which we have previously mentioned, was already a step in this direction. We must remember that the way electronic signals reach the earth in radio broadcasting of pictures is much more suited to a computer than are photographic pictures. Maybe the future does lie in radio broadcasting of pictures. However, there is still a lot of research to be done in the field of evaluation.

Doubtlessly the military has already made some progress which is still being kept secret for the present. We only need to remember that the most advanced computer that will automatically record the differences between two pictures is used by the military. It is maintained by the Control Data Corporation in Minneapolis, and it takes only one half hour to compare two pictures. In this way, the movements of troops are automatically registered on satellite pictures.

When the ERTS satellites were being planned, all of these considerations were still secondary. How they would equip the satellites was of primary importance. Over the years, they have found a solution which indeed is not the optimal one, but which does have quite useful results. The main pieces of equipment used in ERTS are known by the names RBV and MSS.

The abbreviation RBV (Return Beam Vidicom) denotes a television camera system which photographs the earth in three wavelengths. In the blue-green range, bodies of water are most transparent, so that land under water can be observed quite well. In the red range the various agricultural areas are best distinguished, while the near infrared range helps in differentiating land and water masses.

The pictures from the television cameras are sent back to the earth just as automatically as the pictures from the multispectral scanner (MSS), a device which can scan the earth line by line. This scanning takes place in four wavelength areas, three in the optical spectrum and the fourth in the near infrared. Neither far infrared radiation nor radar observation is done within the framework of the ERTS program.

The ERTS satellites have advanced in construction beyond the successful Nimbus satellites. In addition to the above systems, they are equipped with a mechanism for data collection. During their flight over the earth they can regularly store information about thickness of snow cover, rainfall, quality of water, and seismic activity in remote regions. All of the data are sent to one of the ground stations located in Fairbanks, Alaska; Goldstone, California; Greenbelt, Maryland; or Prince Albert, Saskatchewan. Another station in Brazil is currently under construction. Beyond this, there are plans to equip the Apollo ground stations in Spain and Australia to receive ERTS pictures. In this way, the pictures will not have to make a detour over America in order to reach the users.

The orbit of the ERTS satellites was chosen so that they could take all of their pictures of the earth at the same local time, about 9:30 A.M. (sun synchronization), and so that they could photograph the same area every eighteen days. Through this special sun synchronization they have been able to assure that the earth is always illuminated in the same way on all their pictures. At 9:30 the shadow effect is optimal for most purposes. With the eighteen day period, scientists are trying to test how well satellites can pick up changes on the earth's surface – changes such as the gradual ripening of grain or the progress of plant diseases.

The value of such a periodicity was first tested with NASA planes of the U2 type (the same ones whose espionage missions have often caused a furor in other countries.) The planes have been flying over five large ERTS test areas in the United States since August 31, 1971: the Feather River area in northeast California, the areas around San Francisco and Los Angeles, a desert region in Arizona, and the Chesapeake Bay area. They have been flying over these areas regularly at an altitude of about twelve miles, and photographing in an eighteen day cycle.

ERTS 1 has been following this rhythm since July 23, 1972. Every day it sends back a mass of information, which is then evaluated by about 330 scientists throughout the world. Fifty of these scientists are active in twenty European countries; some of them work in Berlin, Frankfurt, Hannover, Heidelberg, and Munich. Most of them are interested in geological problems, followed by ecology and the environment, agriculture, forestry, oceanography, geography, cartography, demography and meteorology.

All these scientists place the most varied demands on the

quality of the pictures. The photographs must be almost geometrically perfect, so that any partial photographs of the same area taken in different wavelengths can be laid over each other. The degree of brightness in these pictures has to be extremely stable, so that it will be possible to evaluate pictures of one and the same area taken during different flights. Finally, in some cases it is important to exactly determine the coordinates of the areas flown over, a process which is actually only possible in those pictures which cover the United States. Determining the coordinates is done with the help of computers. For example, the computers automatically check whether the crosses marked on pictures taken with television cameras are correctly aligned with each other. If they are not, it means that something in the camera mechanism has moved. The corrected pictures are then evaluated based on this information. The computer again automatically seeks out in the pictures several ground control points which the U.S. Geological Survey had determined before the launch of ERTS 1. Six thousand of these points are scattered over the U.S., and every picture is supposed to show at least twenty of them. Now, with the help of exact coordinates, the computer can evaluate the pictures.

This description of the ERTS program may seem too detailed, but it does show us one thing very distinctly: although earth observation satellites photograph the earth just like weather satellites, they also have much greater capabilities. Only a few years ago we did not have the technology necessary to construct them. Meanwhile we have reached a turning point, and can clearly see how fast space exploration has developed in the last few years. One can criticize the American moon landings, but people should not complain too much. We tend to forget too easily that space exploration finally turns back to the earth.

We can also observe this in manned exploration. A typical example is the experimental packet for earth observation which contributed so much of value to the Skylab mission. When the first Skylab crew set out from the space station to start their trip back to earth on June 22, 1973, they had made 16,765 pictures with three ERP cameras. In addition, they had 45,200 feet of tape recordings containing the information from the three ERP sensors. The total amount of data the astronauts brought back from the flight, including the other Skylab experiments, weighed more than the entire Mercury capsule and John Glenn put together!

We can see from the size of the space station just how great are the advances made in manned space exploration. The living and laboratory rooms alone are as large as a one-family house. For their daily life the astronauts have an area of 9,500 cubic feet, which is more than forty times the room in an Apollo capsule.

Having described the ERTS program in such detail, we will only briefly mention Skylab. We must, however, also say a few words about the experimental package for earth observation. EREP, as we have already said, made pictures of the earth which have a much greater resolution power than the ERTS pictures. One of the main reasons for this is that in a manned space flight, they can take along much more sophisticated equipment.

Part of EREP consists of two photographic camera systems whose pictures were not sent back to the earth during the flight, but which were brought back by the astronauts as film material. One system, S 190 A, has four cameras which photograph the earth simultaneously in four different wavelengths. Thus it corresponds to the equipment tested on Apollo 9. The second system, S 190 B, has only one camera specifically intended for cartography.

Experiment S 191, an infrared spectrometer, is probably the most important part of EREP. With it they can take continuous measurements in a broad wavelength spectrum (not just separate measurements in isolated, narrow wavelength) and get much more exact information. This they need especially for the study of the ground structure of surveyed areas, since fingerprints of various soils are quite similar. Geology and other disciplines connected with it profit most from this experiment. While the measurements are being made, the spectrometer has to be aimed at the target; it is not possible to use it in unmanned satellites at present.

EREP also has a multispectral scanner (S 192) which serves the same purpose as the one on the ERTS satellites. (The ERTS scanner is considerably more valuable since it works in a total of thirteen different wavelengths.) In addition, EREP has a microwave radiometer/scatterometer and altimeter (S 193), as well as an L-band radiometer (S 194), both of which carry out measurements in radio wavelengths. The first of these two pieces of equipment is active, that is, it sends out radar waves whose reflections can then be received. The scientists hope for great profit from this equipment especially in investigating the broad expanses of the ocean.

These brief remarks should suffice to cover the topic of EREP. Because of its importance, we really could have given it much more space here, but we do not want to put pure technology too much in the foreground. Earth observation under the EROS program has already had so many surprising results that we should now give this program our attention.

11. The White Spots on Our Earth

The Real Problems of Man

Anyone staying in Lima, Peru a few years ago who went by the palace of the Peruvian President Belaunde could have made a remarkable discovery: on the wall of the palace hung a large picture mosaic showing the entire country at a single glance. As President Belaunde proudly explained to every visitor, this was the very first small-scale map of the Peruvian Andes and of the coastal regions.

This map demonstrates very impressively just how productive photography from outer space can be. A similar charting of the country based on classical field work would have taken decades, if indeed it were at all possible considering the country's lack of roads. Naturally they could have mapped the country using photography from aircraft but simple calculations show that they would have needed 1,000 photographs, necessitating about 50 reconnaissance flights. The entire project would have taken at least a year, and the results would have been a map inferior to the one on the palace wall.

This map was the result of space exploration. In 1966, Gemini 9 flew over the country many times. Eleven photographs of Peru were then joined together to form a picture mosaic, which encompassed a surface area of about 385,000 square miles. The cost of production was minimal, being only a fraction of a penny per square mile. This low price is due to the fact that the map is primarily geological. The main work involved removing distortions from the pictures. At that time this was necessary primarily because space ships were very seldom correctly stabilized. Removing the distortion was quite successful, so that this picture mosaic shows all the surface details in their correct geographical position. This was also the first attempt by the American Geological Research Office to assemble pictures from outer space into a mosaic covering a large area.

Peru now has the use of this map which exhibits the most varied details. The geology of the land can be clearly seen, and many geographical and hydrological characteristics which were unknown prior to this stand out very sharply. Thus, for the first time, Peru is able to implement large-scale development of the country which was previously practically impossible because of the lack of knowledge concerning the conditions which predominate there.

Peru is not the only country facing this problem. A look at the atlas is misleading. Very few areas on the earth are actually known in detail. We estimate that only 2.7% of the earth's surface is mapped correctly. More or less detailed land maps do exist for three quarters of the inhabited earth, but the quality is often deplorable. As satellite pictures have proven, at least one-third of them are totally out of date. Even maps of the United States do not always agree with reality. In Nevada, for example, satellite pictures showed clearly and distinctly the presence of a large volcanic crater which up until then had been unknown.

If objects which have existed for hundreds of years are not correctly perceived and recorded, it is understandable that maps do not keep step with reality. Especially in Europe and North America, development goes on so quickly that geographers are faced with an almost insoluble problem. When they have finally finished a detailed map of a fairly large area, they are forced to realize, to their chagrin, that they have included features which are already out of date. Nature usually wins this race against time.

The problems which arise in developing countries are even more troublesome. Maps for these areas are not even approximately reliable. It is precisely here, however, that they are most essential: no country in the world can be developed without exact planning. How are we to know where to build streets and bridges, where to look for new sources of raw material, where to fence off fields, and which areas are suited for reforestation, if we do not even know the land?

Not long ago these problems became a public concern. At a convention, Brazilian scientists reported a discovery which was so fantastic that nobody would have believed it, if the Brazilians had not proven it beyond a shadow of a doubt. The scientists had taken a picture from the satellite ERTS 1 showing the Amazon area, and compared it with the best existing maps of the same region. The general director of the Brazilian Space Exploration Office, Fernando de Mendonca, discussed the results of this comparison as follows: "We have determined that in several places the world's largest river, the Amazon, is drawn on the maps with a factor of error exceeding twenty miles. Several tributaries of the Amazon flow in entirely different directions than previously thought. In some cases, the differences approach 90°. We have also made out islands which are not even shown on our maps!"

The discovery would have been exciting enough, but not as dramatic, if Brazil had not recently opened up a part of the new Amazon super highway. An enormous number of bridges had been built in the area in question, and the pictures from space now indicated that several dozen of them were completely superfluous. If these pictures had been available earlier, a lot of money could have been saved.

Especially in Brazil observation of the earth from outer space has proven to be very advantageous. Much of the entire Amazon area is still unexplored. One of the tasks of the second Skylab mission, for example, consisted of observing a 375 mile long river, west of the jungle city Manaus. This tributary of the Amazon seems to contain valuable gold, diamond, iron ore, and tin deposits. Despite its length, the river was only recently discovered, and it still does not have a name.

We should not be too quick to claim that mapping of

these countries is a problem that does not affect us. On the contrary, we should hope that all developing countries are opened up as quickly as possible. We need business partners for tomorrow. If the western world wishes to maintain its standard of living and to raise that of the developing countries, then it must rely on an expansion of world-wide commerce. This can only be done by making the developing countries our concern.

Mankind must proceed very carefully. It is easy to raise productivity and to expand world-wide commerce, but if we are careless in this respect, one day there could be a great collapse. Prophets of doom predict that this will happen by the year 2050 at the latest.

This prognosis may be a bit too hasty. At present, however, the world population is increasing by 190,000 persons per day. A decrease in this growth rate is not now in sight; for ethical and religious reasons, global birth control is presently not open to discussion. It breaks down especially because of the backwardness of many peoples whose sole joy seems to be the birth of children. The incorrect use of birth control pills mentioned previously should give us food for thought.

The increase in world population – statistics mention 6.5 billion people in the year 2000 – combined with maintaining the standard of living in the western world, or increasing it in the developing countries, will have a fatal by-product. If we do not come up with an answer as quickly as possible, our food, as well as our sources of raw materials and energy, will soon be exhausted. The consequences of this are only too familiar, in view of the recent energy crisis resulting from the oil embargo by Arabian countries.

There are already danger signals. At the time of the first Sputnik in 1957 an acute lack of protein was widespread in many parts of the world. In its yearly report 1963–64, the UN organization for nutrition and agriculture determined that, on the average, the per capita production of food in the whole

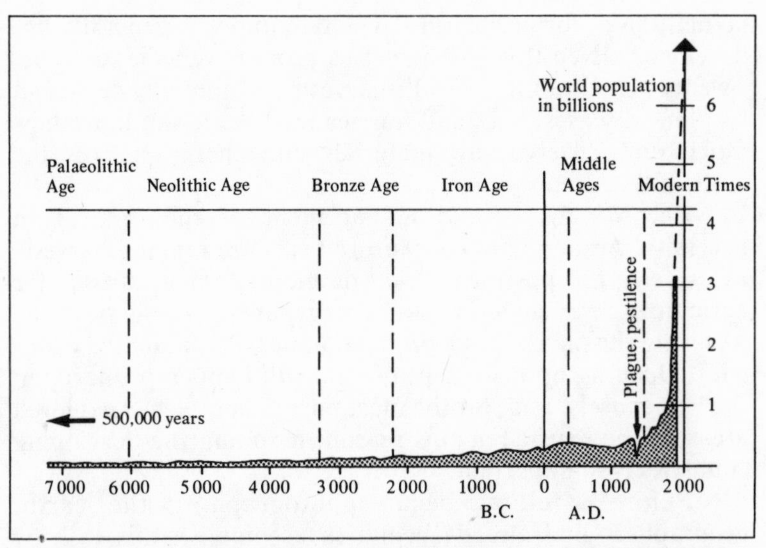

If the present development continues, and mankind increases at the same rate as previously, about 6.5 billion people will inhabit the earth in the year 2,000.

world had not risen since 1958. Nobody actually knows to what extent the raw material and energy sources of the earth have been depleted.

On the other hand we are faced with mankind's growing needs. By the year 2000 we will have to have twice as much food as today. The need for raw materials will increase five-fold, and the need for energy ten-fold! The question is: where do we get all this without robbing ourselves? We can be sure that the earth is not getting any larger. Its living zone, the biosphere, is, and remains limited. As a matter of fact, it is being shrunk artificially, because people do not stop soiling their environment.

Mankind seems to be stuck in a hopeless situation, but it is not yet the eleventh hour, as pessimists claim. If we look more closely, we will see that only 30% of all arable land is actually used for agriculture. The remainder is generally undeveloped. Even this 30% is not as productive as it could be. Twenty percent of the world's harvests are annually destroyed by plant diseases, hail, and hurricanes. We do not know how much undiscovered raw materials and energy sources the earth still conceals.

Accordingly, careful management of the earth is a necessity. Among our other future tasks are reaping harvests as safely as possible, and developing new areas for agriculture, raw materials and energy, as quickly as possible. Assuring the harvests is a problem in heavily populated countries. Opening up new strips of land will happen primarily in the less densely, and, for the most part, incompletely explored areas of the earth. For this reason, mapping the developing countries is so important.

Before satellites began photographing the earth, geographers had already begun to accomplish this task by using airplanes. Airplanes flew over tracts of land, and bit by bit picture mosaics of larger areas resulted. They showed that air reconnaissance offered a means of drawing up maps of remote regions much quicker than had been possible with traditional methods of field work.

However, it gradually became obvious that even air reconnaissance was limited. For the solution of the problems facing us, this method is still much too slow. For example, we would need 1.5 million airplane pictures to encompass the entire United States. Shooting these pictures would take no less than ten years, and the cost would run to $12 million!

The situation changes with the use of satellites. We now could get by with about 400 pictures that could be shot in only seventeen days. The cost would amount to $750,000. Naturally these are only estimates. The number of pictures necessary depends essentially on the altitude at which the airplane or satellite flies, and the costs and amount of time involved vary accordingly. However, the relative magnitude of the fitures cannot be questioned.

We also cannot overlook the fact that satellite pictures are generally much more valuable than airplane pictures. During airplane reconnaissance, the position of the sun changes continually. This means that the effect of the shadows also changes, so that composition is extremely difficult. Satellites, on the other hand can fly synchronous with the sun, so that the conditions of illumination are always the same.

Furthermore, the geometry of small frame satellite pictures is better than that of airplane pictures. For airplane reconnaissance pictures to be worth the effort, they must be taken with relatively wide-angle lenses. Thus they show considerable distortion on the edges, which often cannot easily be removed. Since satellite pictures scarcely have this problem at all, map production is speeded up – one more advantage of satellite photography.

A picture from Apollo 9 taken over the Phoenix, Arizona area proved that these considerations are not just theoretical. This picture was compared with a map drawn from photographs made with a precision surveyor's camera. At first the map and the picture seemed to agree completely.

An enlargement of the picture, however, revealed a number of details which did not agree with those on the map. Close examination showed that it was not the Apollo picture that was wrong, but rather the map, even though the resolution of the surveying camera was better by two orders of magnitude that that used for the Apollo picture!

Generally speaking, one was able to state with relative certainty even before using the technological earth reconnaissance satellite ERTS 1 that, in comparison with other methods of cartography, satellites would show advantages in shooting broad areas, and in examining remote regions. Furthermore – and this we will return to later – it would have the advantage that they could photograph one and the same area repeatedly.

This prediction was confirmed in practice. Within the

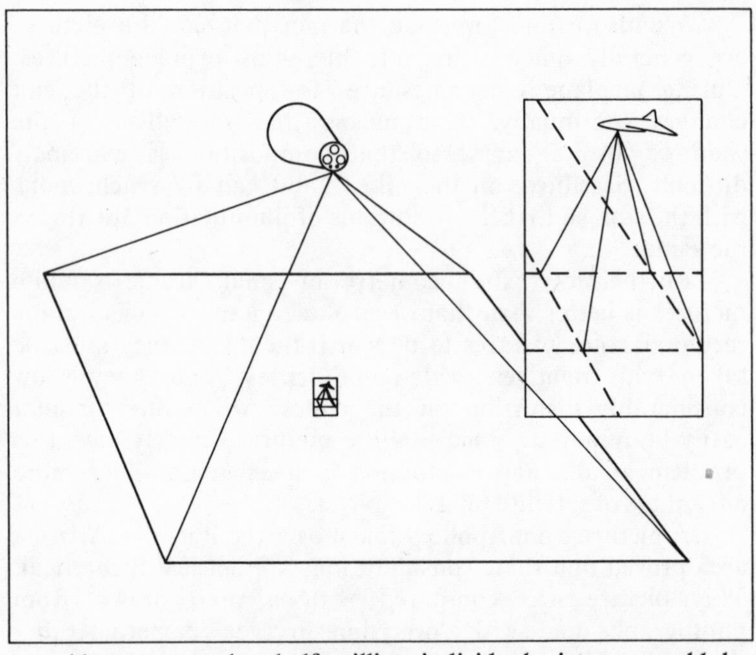

About one and a half million individual pictures would be necessary to map the United States using airplane photography. Satellite pictures can cover the same area in about 400 pictures. For purposes of clarity, the altitudes in this sketch are not drawn to scale.

first year, this satellite took about 70,000 individual pictures, including 75% of the land mases of our earth. Without exception, the areas not included were in the Soviet Union and China. This is due to the fact that ERTS 1 is still a technological satellite. Outside the United States its photographic equipment is turned on only according to the wishes of the scientists who are taking part in the project: there are none in those two countries.

Worldwide Cartography

The charting of the earth stands at a turning point; we

are starting down a new road. The time is past when isolated successes had to be played up to demonstrate the value of satellites in cartography. It is a fact that pictures from the Nimbus satellites have already led to corrections in our maps. These corrections caused scientists to predict that a worldwide program for photographing the earth in detail from outer space could realistically be carried out with a cost to returns ratio of 1:20. That is, every invested dollar would produce a gain of twenty dollars.

What kind of results did the satellites achieve? Mainly corrections to maps of the most impenetrable and remote areas of the world. With the help of the Nimbus satellites, many atolls in the South Seas were resurveyed. In several cases it turned out that their location and shape were incorrectly indicated in the existing atlases. Satellite pictures were also used, along with other data, to produce new maps of several areas in Tibet and China.

Corrections were especially necessary in the Antarctic. In one of the larger projects, 300 Nimbus 1 photographs of this area were analyzed very closely. They showed that the ice floe area had to be revised. The floes in the Filchner ice floe area, in the Weddel Sea, and in the area of the Princess Martha coast followed a completely different course than the maps showed.

Even more dismaying was the realization that there was a mountain on the maps of the Antarctic which could not be found on the Nimbus pictures. This would not have surprised anyone, if just some small hill had been involved. However, what they were looking for was Mt. Siple, 9,700 feet high, a mountain which airplane pilots used as a navigation aid. What kind of navigation that must have been! Finally they located the mountain on the pictures, two degrees, or 45 miles, further west than was indicated on the maps. This mistake has now been corrected; the mountain has been "moved".

Another change in maps of Antarctica concerned the location of a mountain group. Two expeditions had sighted it in the Kohler Range, but, unfortunately, at different places. Of course only one mountain group existed. Using pictures from Nimbus 1, the mountains were finally correctly located on the maps.

Yet, as we have said, these were isolated results. We really could not expect otherwise, because the power of resolution was not particularly good for surveying, and because there were geometric errors in the weather satellite pictures – two factors which do not particularly bother meteorologists. Basically the results obtained with photography from outer space from the manned Gemini and Apollo flights can also be considered isolated. However, the situation was somewhat different. These pictures showed not only unique new details, but for the most part many features which were totally unknown until then. They revealed new lakes and volcanic formations discovered in Persia, and helped in eliminating mistakes in the maps of the Marshall Islands in the Pacific. The only limitation was that the astronauts could only photograph the earth fragmentarily during the flight.

The fact that in a few cases connected photographic maps of fairly large areas were produced does not change the general picture at all. One example of such a large map is that of Peru which we have already mentioned. Another example is a mosaic covering 270,000 square miles, from the Gulf of California to central Texas. It shows many details concerning the geology, water, and ground distribution, and so on, which are scarcely recognizable on conventional maps.

We should also not forget to mention that photographs from espionage satellites are also used to produce new maps, although hardly a word about this ever reaches the public. The military generally show no interest in revealing how they make their maps. Occasionally remarks do filter out from the jungle of secrecy, as was the case, for example, with the short statement that American Army satellite pictures were used in making maps of parts of Borneo.

The introduction of earth reconnaissance satellites should mean that these individual results will become one worldwide, lasting achievement. Many countries in the world have been looking forward to this for a long time. They are waiting for the day when they will finally get the all-inclusive survey of their area which is so absolutely necessary for extended development. For some countries this charting from outer space will only be the dot on the i. Not many people know that in the last few years wide areas in South America have been mapped from the air.

South Americans have completely given up traditional aerial photography, since it is not ideal for large-scale projects. They have been using radar charting from the air. This is sufficient for many purposes, especially as this method does not show the distortions of aerial photography. In this way, the South Americans were able to fill in the gap until the introduction of photography from outer space, and thus to map their regions in a practical way.

One of the largest radar reconnaissance projects was begun in mid 1971 in the Orinoco Valley in Venezuela. They were looking for mineral deposits, areas favorable to agriculture, and routes suitable for future highways. AERO Service Corporation of Litton Industries won the contract, and carried out its task to the satisfaction of their employer. It examined 3,100 square miles of surface area every hour. Neither fog nor clouds hindered the work. Using the Transit Navigation Satellite System they succeeded, through a complicated process, in reducing the error in the maps to about 33 feet.

The contract was soon extended. Brazil also joined the project. Within a year hundreds of thousands of square miles of land in Venezuela and Brazil were mapped in this way! This confirmed the results obtained with radar reconnaissance in Columbia, as well as in Indonesia. Under these circumstances it is not surprising that the South Americans are eagerly waiting for the results of earth reconnaissance from outer space. The combination of various reconnaissance methods is the best way to attain optimal results in any case. We must also not overlook the fact that ERTS 1 photographs the earth in wave length ranges that give much additional information not available in radar pictures.

Brazil alone has allocated ten million dollars for the reconnaissance of its country. Through the combined use of satellites – for the moment, this means ERTS 1 –, airplanes, and ground troops, they are expecting valuable results in charting minerals and water, in highway planning, in agricultural and industrial development, as well as in the supervision of seasonal vegetation conditions. Two hundred Brazilian scientists have taken courses in the United States to prepare for this project. Their efforts will surely be repaid. The new cartography of the Amazon was one of their first successes.

Likewise other countries in South America use the technological earth reconnaissance satellite ERTS 1 in combination with other methods of observation. The list of countries which have asked NASA and the American Geological Research Office for support includes Venezuela, Argentina, Chile, Peru, and Bolivia. There are also, of course, many non-American countries which have shown interest in the pictures from ERTS 1. Bangladesh, for example, has invested $175,000 in a program which mainly includes studies for land use, water resources, and crops.

It will be quite a while before all the projects planned are carried out and evaluated. Maybe the knowledge gathered will then be sufficient to establish a truly operational earth reconnaissance satellite system, following in the tracks of ERTS and Skylab. The point will then have been reached when NASA will withdraw and hand over the routine work to a separate organization, just as it did with the communications and weather satellites.

From the Earth's History Book

We might wonder what importance the American earth

reconnaissance program has for a country such as Germany, where every area has already been exhaustively mapped. For the moment even regular observation of fields and forests from outer space promises no great profit, since the resolution on the pictures from ERTS 1 is far too inadequate for this. In Germany, division of land is not planned on so large a scale as in the United States. The system of parcelling land goes merrily on, and it is practically impossible to get any further details from outer space.

Nonetheless, ERTS 1 has given the Germans new knowledge about their country. There are structures that are so extensive that by using transitional methods (including aerial photography) they remain for the most part unexplored. What we mean here are the geological structures important in understanding our planet.

Folds, dislocations, ruptures, and lineaments often stretch over dozens of miles across the land. Although they are partially accessible to geological field work, they are, whether because of vegetation or some other phenomenon, partially hidden from the scientists' eyes. Our knowledge of them is therefore fragmentary.

It often happens that there are very good maps of neighboring regions that clearly show geological formations. However, they were produced by different scientists, one emphasized this phenomenon, and the other that one. Every geological map bears the stamp of its producer; it is practically impossible to combine maps so that supraregional structures become visible.

We thought we had an exact knowledge of the Alps until the first ERTS photographs arrived in central Europe in 1972. Every corner of the range was known, every mountain peak investigated. But to our surprise these first pictures showed arrow straight lines whose existence no one had suspected. Comparison with the geological maps of the area revealed the reason for our previous ignorance: various sections of the lineaments had already been surveyed, but between them there were sections concealed by other phenomena, and no one had thought of comparing the surveyed areas with each other.

It will be a while before the scientists, using satellite pictures, can come up with a new history of the Alps – as it is revealed in these structures. What brought about the formation of the mountain range, what happened when the mountains arose and separated Italy from the rest of Europe by a steep barrier? We still cannot provide the ultimate answers; extensive field work is necessary to clear up these questions.

If pictures from outer space reveal new geological details about central Europe, how many more details will they reveal in other, remoter areas of the world? As a matter of fact, it was felt that the first practical application of such pictures was in the field of geology. This was revealed by the American Mercury flights of 1962–1963. The snapshots the Mercury astronauts made of the earth were quite primitive compared with the pictures obtained in the Gemini and Apollo projects, but for the geologists they were sensational. They showed a number of details with previously unattained clarity which were enormously important for knowledge of our earth.

The scientists began working industriously, especially on dry regions. It is not a chance that these areas were photographed much more frequently in the beginning than were others, as dryness enhances the clarity of photographs. There is little water vapor in dry atmospheres, and the result is a brilliance of color never attainable in damper regions. A further factor was that the Mercury capsules, because of their orbits, only flew over areas of limited width where the very dry regions are most common.

The first geological map composed from Mercury 4 pictures was soon ready. It included the entire Sahara area. The scientists at McGill University in Canada who drew it up had included a multitude of details, and the information content was so great that, with one blow, it made all earlier geological surveys of the Sahara appear fragmentary. One side effect of this and other evaluations of the Mercury photographs was that much more stress was placed on earth photography in the later manned space flight programs.

Based on space photography, geological maps of other regions on the earth were also produced in the course of time. Again and again surprising new details became visible. For example, a geological mapping of Mexico was made from Gemini pictures with a scale of 1:250,000. It showed a young volcanic area with a diameter of twelve miles on the border between New Mexico and Chihuahua. Just one year earlier a newly published map had designated a part of this very region as extremely old. There was no reference to young volcanic activity.

Determination of volcanic areas does not seem to be very easy. It is therefore not surprising that scientists have argued for centuries whether lunar craters are volcanic in origin, or are the result of meteorite impact. Let us take as an example a large, circular crater structure in Mauritania. Without hesitation, the scientists had attributed this crater to the impact of a great meteor. There were no indications inconsistant with this idea, not even in airplane picture mosaics. It was space photography which first showed the entire area, and proved that this idea was nonsensical. Today we know beyond a doubt that the Er-Richat-Structure is volcanic in origin.

One of the special tasks of the earth reconnaissance program is to study the volcanic areas of our earth. The practical application of these studies is considered of value. That is, the studies are designed to collect knowledge which will make it possible in the not too distant future to judge the activity of volcanoes and to predict eruptions. A beginning has already been made. Weather satellites using infrared sensitive equipment that registers the earth's heat radiation, have succeeded in measuring the increase in temperature in the interior of volcanoes before an eruption occurs. Such measurements however, are the exception.

It is difficult to estimate the practical value of these measurements, since there are still insufficient observations on the temperature conditions within volcanoes. Infrared investigations of Mt. Kilauea, Hawaii, one of the most active volcanoes in the world, did not show until 1965 that a number of heat veins were concealed underneath it. Differences in temperature which can often be considerable continually occur in these veins and volcanic chimneys. Where is the border line between erupting and not erupting? Only the future will answer this question, and then heat observations will be applied in predictions.

ERTS 1 does not have any infrared sensors that react to heat radiation. Instead the scientists developed something different for this satellite. Naturally they started with the assumption that the warning of an eruption often has to be given very quickly, and thus must not take too much time. They also took into account the fact that the satellite possessed data collecting equipment, and this provided the answer: ERTS 1 had to assume the role of a communications satellite.

Sensitive seismometers registering the tiniest movement in the earth were set up in various areas of volcanic activity. As soon as a volcano becomes active, begins to live, the earth starts to tremble. This means that the seismometers' needles begin to quiver. ERTS 1 records these measurements when it flies over the area, and shortly thereafter sends them to the gound station at the Goddard Space Flight Center in Greenbelt, Md. From here a report is immediately sent by telephone or teletype to the Menlo Park Center of the American Geological Research Office. Moments later a warning is on its way to the threatened area.

On February 23, 1973 the system underwent its baptism by fire. Just six days previously one of the seismometer stations had been installed on the cliffs of the volcano Fuego in Guatemala. Another station already existed twenty miles away. The equipment functioned normally for four days, reporting a constant, but not disturbing, volcanic activity. The earth trembled five times daily. Suddenly the activity increased by leaps and bounds, and within two days hell had broken loose in this previously fairly quiet area. On February 23 the volcano erupted. Fire, lava, and ashes were spewn from its throat, but nobody was surprised. Based on the reports that the satellite had broadcast, this had to happen.

Modern Treasure Hunts

Investigating the volcanic regions is closely connected with the question of our earth's history. We have repeatedly seen that many lineaments are associated with seismic activity, and that where the earth quakes, volcanoes are not far away. Our world is not calm at all. In central Europe this was all too often ignored. Inside, the earth is simmering. Or is it a raging inferno? We can get a glimpse at the interior dynamics through the weak zones of the earth.

How many disasters our planet's convulsions have already caused, and how many more they are still causing, can be clearly seen in the statistics of great natural catastrophes. Again and again people lose their lives in volcanic eruptions or earthquakes; entire cities are destroyed and leveled. The destruction of Pompei was not a unique incident. Nonetheless we do have to thank the earth's "vitality" for good things as well. After all, without it, we would not have the sources of raw materials we so desire, and the world would only be half as beautiful. Although the origin of minerals and fuels is still not completely explained, we do know that a wide variety of deposits are found in greater quantity along lineaments, especially where these intersect.

Such deposits are usually invisible from the outside. Only occasionally do they leave clearly discernable traces. Copper deposits, for example, can produce a measurable increase in temperature, because of the heat of oxydation. The melting snow in the spring reveals their location. Generally, however, the task of the prospectors, the modern treasure hunters, is not easy. They have to study the geology of a country in great detail. As soon as they have discovered certain geological structures, they send out the field troops. They must then by means of tedious manual work, check whether minerals or fuels are present in the earth. Prospectors have been working with aerial reconnaissance for years. They climb into an airplane, fly over and photograph an unknown area, and finally sit down at their desks to evaluate the pictures. Naturally, they are looking for the geological structures near which sources of raw materials often occur, and not for the raw materials themselves. This is still left up to field work. Nevertheless, aerial reconnaissance has made prospecting far more profitable. Work which used to take weeks and months is now carried on relatively quickly, and with much greater success.

When the first pictures from outer space were published in the sixties, the technologists immediately grabbed onto them. They wanted to check whether prospecting could be done more effectively with the help of these pictures. This was to be expected, because of the special value which this photography possesses for the geologists. The pictures did indeed prove to be of value. A series of photographs revealed structures which positively reeked of deposits, and in many cases investigation at the source showed the correctness of that assumption.

The first concrete indications of the possibilities for "treasure hunting" from outer space were provided by the pictures the Gemini 4 astronauts took in 1965. They show many details which clearly indicate the presence of oil and gas deposits in North Africa. Up until then these had not been suspected. Soon after this, expectations of such deposits increased, when the pictures from Gemini 6 were evaluated. Signs of raw material deposits of almost unlimited proportions were found in middle Africa, in the mountain area near Air in Nigeria, and in the uncharted mountain regions on Lake Chad. Linear structures and circular shapes in the country's geology promised the people immeasurable riches.

It will not be long before Nigeria decides to exploit its wealth. We can only hope that it will do this intelligently. If it does, it will be among the first developing nations to undergo an economic development made possible by space flight – a development considered impossible only a short time ago. The pictures from Gemini 6 should console us too. What happened in Nigeria will be repeated elsewhere. We no longer need worry so much that the earth's sources of raw materials and energy are being depleted. Nobody has any idea how much is still waiting to be discovered.

Because the pictures from Gemini 6 were so sensational, it was a foregone conclusion that industry would soon attempt to join this enticing business. The big run on pictures from outer space began when the space ship Apollo 7 returned to the earth with its booty. Within a very brief time, industry got hold of these promising pictures, and the experts feverishly began looking for signs pointing to previously unknown, exploitable sources of raw materials.

No one likes to talk about the time that NASA really got itself into hot water. Its broad-mindedness caused problems, and the reason is obvious. When NASA released the pictures, whoever evaluated them first had the best chance of obtaining the rights to large areas, where no one had previously suspected hidden treasures. Naturally this was primarily the Americans. Did this drive the developing nations into a further, oppressive dependence on the Americans?

The Apollo 7 pictures contained many such indications of minerals. Geological structures pointed mainly to oil – oil in southern Morocco, in Algeria, and in Australia. Australia is a good example of what happened at that time. It did not take long before test drillings appeared on the Great Barrier Reef, and soon after this the entire oil-bearing area was divided up for plundering!

Such occurrences cause bad feelings. The English particularly reproached the Americans with using unfair means to gain advantages in the oil business. Nobody had anything against using pictures from outer space. Quite the contrary: this possibility was considered very promising, and an indication of the future. What was intolerable was that the Americans cemented their advantageous position in the economic world in this way. NASA acted upon these reproaches. After this, it always held onto some of the pictures that the manned missions brought back, and did not make them public.

Treasure hunting from outer space has been discussed repeatedly. Is it legal to photograph other countries from space without their specific permission? Is it legal to gain advantages for one's self from the information that these pictures contain? Ultimately, the discussion involved a much wider scope of questions than just treasure hunting alone. For example, it could soon be possible to estimate the size of harvests using infrared photographs. Whoever possesses this knowledge can intentionally raise or lower the prices on the world market. The Military aspects (espionage) should also not be ignored.

NASA bases its claim that it be allowed to photograph the entire earth from outer space without limitations on a 1961 UN resolution. This resolution says that outer space is open to any kind of research. Since NASA turns its projects over to other institutions to be carried out, it is in the clear. This does not mean, however, that the problem is solved. At this point we should mention the statement of a technologist who seriously suggested that the government leave the pioneering work in the area of earth reconnaissance to non-government institutions in order to avoid political complications. Is this a solution?

NASA is to be commended now that it publishes its pictures and makes them equally available to all interested parties throughout the world at the same time. This means that the scientists who are involved in evaluating the pictures from Skylab and ERTS 1 do not get the pictures from the EROS data center in Sioux Falls until everyone in the world can get them.

Furthermore, NASA points to the fact that, generally speaking, these pictures from outer space alone are not sufficient to find new sources of raw materials. Field work is necessary, and in a foreign country this work can only be done with the permission of the local government. The question is whether this argument is relevant in every case; we are not able to give the solution to this problem here. One must agree that NASA is really trying to help other countries develop, especially those in the third world. Indeed it is also trying to restrain dangerous competition. For example, it admits that the tributaries of the Amazon in Brazil apparently contain much gold – at least the ERTS pictures would make us think this – but it does not reveal the exact location of these tributaries.

In order to judge NASA's position, we will give an example of how treasure hunting using photographs from outer space actually proceeds. Martin Marietta Corporation is involved in the development of EREP, and thus also in the practical use of remote reconnaissance equipment. A short while ago, it got a contract from its sister organization, the Aggregates Division, to look for new deposits of limestone in Florida. Because of a boom in the building industries, this material had become scarce, and the nearest known large sources of limestone lay too far away to be transported economically.

First the technicians of the Martin Marietta Corporation obtained an Apollo 9 photograph from the EROS data center in Sioux Falls. They also got a few airplane photographs from the Technical Applications Center in Albequerque, New Mexico, the data bank for the photographs of the Gemini and Apollo flights. They claimed that these pictures were availagle to everyone. The Apollo 9 picture showed a number of regions which could be deposits of limestone. These regions were then examined more closely on the airplane photographs.

A total of nine places were discovered that seemed to have limestone deposits which could be mined on or near the surface. The field troops then moved out; in all nine cases, they were able to prove that this was correct. At that point they had used only 10% of the time which would have been required to obtain the same result if traditional methods were used. In the last phase, Martin Marietta determined who owned the land, and obtained permission for test borings. With this step, it had completed its task satisfactorily. (In a similar way, the corporation discovered new copper and molybdenum deposits for a prospecting company in Colorado and Arizona.)

This then is the procedure one must follow in order to apply photography from outer space in the search for minerals. It will probably be used many times in the future. Through the evaluation of such pictures, a number of countries have come to realize that their territory also looks promising. In Alaska and in western Texas there seem to be enormous petroleum deposits which were previously unknown. Many different kinds of ground deposits are suspected in central Asia, from Afghanistan, across the southern parts of the Soviet Union and the Hindu Kush, to Pakistan. Strongly colored lakes in northern Argentina and Paraguay are also cause for speculation. Gold is probably contained in the tributaries of the Amazon. This list could go on and on. Prospectors will not have to complain about a lack of work in the next few years.

12. The Need for Good Management

Rosy Times for Agriculture

Not long ago a train rolled across Canada's expanses. All that might have differentiated it from other trains was its cargo: sulphur and gas. This train would never have gone down in history, if it had not had an unfortunate accident. The actual events are difficult to reconstruct. The wheels probably overheated because of long hours of friction. In any case, the cargo suddenly burst into flames, and that was fatal. During the combustion, sulphur dioxide developed, a very lethal plant poison. Some of this gas was released and spread in the air destroying an enormous agricultural area along the railroad tracks.

At the time a secret project was initiated which used ERTS pictures to examine the extent of the damage. Secret! This does not mean that the pictures themselves were secret. Anyone interested who knew where the accident occurred could order them for examination. Only the results of the investigation ended up in a safe; they are of great interest to the military. Again and again entire harvests are destroyed; they are the victims of catastrophes. Fires, storms, and insects continue to cause havoc. Plant diseases also contribute to the destruction of the world's harvests. Just how great a loss the farmers have to suffer in this way can only be estimated. It probably amounts to between $13–20 billion per year.

That would be bad enough, but mankind is increasingly dependent on good harvests every year. There is already too much famine in the world. The situation could be controlled if harvests got better, not worse, and if new farming land were found and cultivated. Elimination of plant diseases through early recognition from outer space, or by any other helpful measure, would produce greater yields. Every dollar invested in this effort could return five dollars profit.

Financial gain alone is not decisive; ultimately we are facing no less a task than assuring the dignified survival of mankind. This will only be possible, if we apply all available techniques. In the United States, agriculture has reached such a high level of productivity that, in spite of large infertile areas, three farmers provide food for one hundred people. The division of fields is planned on such a large scale that the parcels can be recognized from outer space. This means that in this part of the world a precise estimate of the crop yield can be obtained. Traditional methods could indeed be used for this purpose, but they can only succeed now and then. Estimation of larger areas is a precondition, and can only be carried out annually at a tremendous expense in time and money.

It is obvious that there are several obstacles which prevent a worldwide harvest estimate. In most areas of the world, including Europe, the division of fields is not carried out on a large enough scale to make individual fields recognizable on pictures from outer space. We will, however, find the ways and means to refine present observation technology, and perhaps then we will reach this goal. That is what the world needs – a global crop estimate for the future. We need a global land utilization map which tells us what kind of plant or grain types grow best and should therefore be planted in each and every place on the earth, so that we can utilize the soil to the utmost; we need a map which tells us how the crops mature, and what yields we should expect. With this information we could manage to nourish all mankind. In a sensible system, we could compare the available food supplies with the need in various regions (the old song of supply and demand), and then we would obtain optimum results.

Wouldn't it be wonderful if there were no more mountains of excess butter, if no more shiploads of coffee had to be dumped into the sea? In countries where a poor crop is expected we would be able to supply people with food from other regions of the world, before it was too late, before famine had ravaged the population. We could deliver the food well in advance, when the approaching disaster was first recognized.

Is all of this only a dream, or is it really possible to reach this goal? We still have no answer to this question. In order to create a worldwide land utilization map, we must first be able to observe all arable regions of our earth from outer space. It is precisely this problem which we have only partially solved, because, as we have said, the fields the farmers cultivate are only rarely large enough to stand out individually on pictures from outer space. This we must bear in mind.

When the fields are large enough, then the photographs show what is planted on them. Progress has already been made in this direction, although we have not yet reached the limits of what is possible. Photographing the earth in various wave lengths has proved extremely valuable. The characteristics of the various types of vegetation can be easily differentiated, especially in the infrared range where a green leaf reflects, on the average, about 80% of the radiation striking it (as opposed to 20% in the green wave lengths, and 5% in the red wave lengths). Consequently, to distinguish vegetation from outer space is not merely theoretical.

Healthy vegetation stands out clearly; cultivated fields appear in all shades from bright to deep red. Forests are a darker red, with the deciduous trees brighter than the coniferous trees. Even the size of the leaves affects the reflection of radiation. Dying vegetation causes a change in color to blue or bluegreen, and then finally black. Insect damage to plants is revealed as black spots. Since water also appears black, areas with large amounts of ground moisture are darker than dry regions. Settlements and streets exhibit a very characteristic blue, as do all inorganic substances.

Identification of agricultural areas using satellite pictures is a science in itself. The effects must be differentiated. Does dark coloration indicate insect damage, or high moisture content of the ground? Blue – is that dying vegetation, or a building complex? A look at pictures taken in other wave length ranges often helps. This, then is the way to revolutionize the future of agriculture. How far have we actually gotten?

We have come part way. We can only mention a few examples here which are in a sense characteristic. In one case scientists at the University of California at Berkeley examined an ERTS photograph of southern California. It took only a few minutes to shoot this picture and relay it to the ground station, while photographing the same area from the air had taken 250 flight hours. This is only incidental, however, Using the ERTS pictures, the scientists attempted to chart four different types of cultivation which had been previously determined. Their success was enormous. Comparison with the actual values gained in painstaking field work showed that 83% of the cultivated areas had been correctly identified. Mistakes occurred primarily in the case of smaller cultivated areas.

With similar precision the scientists of Purdue University, using an ERTS photograph, determined how much pasture land, how much cultivated land, and how much water existed in a specific part of Texas and Oklahoma. The analysis was so good that it even distinguished different qualities of water. Furthermore, the greatest part of the work was done by computer, so that the time expended in this investigation was only 48 hours! Just a few years ago this would have been unimaginable.

Evaluation of the pictures requires utmost precision. This became obvious not so long ago. Again using an ERTS photograph, the rice fields near Sacramento, California were to be measured as a test. The scientists came up with an area of 15.8 square miles, while the field work run at the same time with this test had shown 18.9. They were not satisfied with this result. Where were the missing three square miles of rice?

The scientists rechecked their data very carefully, and suddenly it dawned on them that there were many straight lines within the fields. These were irrigation ditches and boundaries where no rice grew, a phenomenon very typical of rice fields. The entire measurement was repeated, and what was the result? The rice field grew to 18.5 square miles. The scientists were now satisfied.

Naturally we must take into account that all these results in field identification using ERTS photographs were gained in model fields. These are areas with many extensive single crop fields, and in these circumstances appraisal of the farmland is relatively simple. There is no reason why we should not look optimistically to the future. In spite of the grave limitations presently existing, we will probably one day reach the goal of a worldwide crop estimate.

To be exact we have two more steps to take before we reach this point. The only thing we can actually get with photography from outer space is an estimate of the amount of farmland. Such estimates, incidentally, are already carried out in the United States every five years, by traditional methods. For the future, thanks to the explosive development in space flight technology, it will be economically feasible to go to a yearly cycle.

Only once has the attempt been made to predict a crop yield based on an acreage estimate, and that was only for wheat, rice, and other cereals. The results were so fantastic that there seems to be nothing else in the way of good crop estimates, provided we have an idea of the acreage. The range of error here was only 2–3%! This is well within the expected range of statistical error.

Identification of cultivated areas is doubtless a tremendous service provided by photography from outer space. The decision whether an area should be cultivated, and what crops should be planted is also important for the development of new agricultural regions. In the long run, agriculture has to be expanded considerably. We know that not every soil is suitable for every crop; while one plant needs a wet soil, another needs dry soil. Many areas cannot even be considered for cultivation, because of their high salt content.

Although attempts have already been made in this area, it is difficult to answer the question of optimal cultivation. The tests carried out on the Skylab mission with the experimental packet for earth reconnaissance were intended to clarify the identification of various soils. It will be a while before the experiments have been evaluated sufficiently to have practical significance.

In many cases we can already determine whether an area is being cultivated correctly. This also provides valuable knowledge. It can lead directly to the introduction of numerous relief measures, and this means that the existing fields will then be used to the fullest. The best example of this is the Imperial Valley, south of Lake Salton in California.

The Imperial Valley is artificially irrigated. Large amounts of water are piped from the Colorado River daily, and ingeniously distributed. If proper care were not taken, the water would evaporate under the burning sun; the salts would leach out and turn the area into a desert. Thus, a sensible system of underground runoff pipes has been constructed which make sure that the water is carried away together with the minerals. The result is an exemplary, fruitful farm valley.

Conditions are completely different in the area south of this which belongs to Mexico. The Mexicans have failed to build a similarly effective runoff system. They are now bearing the consequences. The ground is much saltier than in the surrounding land, and the fields produce fewer crops. The difference is so great that various pictures from outer space clearly show the border between Mexico and the United States. Thus, in certain circumstances, we can identify even political boundaries from outer space. Who would have thought that possible?

Centers of Population

Earth reconnaissance satellites will prove much more valuable for agriculture than for geology, because geological processes are long-range. Folds which stretch across the earth's surface last for millions of year. Fields, on the other hand, alter their appearance very quickly. They call for continuous observation. Once the operational earth reconnaissance satellites have been put into use, it will not be long before the earth's entire geology is charted. Then the satellites will be used almost exclusively for observing those structures which are subject to rapid change.

Naturally cultivated areas belong in this group. However, they are only one among many types of land utilization, all of which are equally important. (We will discuss other changing phenomena on the earth's surface later.) It is of course necessary to increase the extent of cultivated areas, but we must proceed with caution. After all, people must live some place, and sufficient space must be reserved for settlements which are laid out sensibly. The forests, whose importance we are only gradually realizing, must not be cut back too severely.

It is very difficult to deal reasonably with all three categories at the same time. It is surprising that until now their interaction has remained so incompletely examined. Evidently, pictures from outer space were first necessary to show man how poorly he has been handling his environment. We suddenly realized that many things were in bad shape. The pictures clearly showed a terribly senseless, destructive settlement on the east coast of the United States which could easily have been avoided. The scientists then decided to test to what extent they could use their earth reconnaissance satellites to keep an eye on such phenomena. Several experts now took up the problem of distinguishing various land utilization types on the photographs.

Using pictures from ERTS 1 they hoped to be able to identify about nine different categories. Detailed studies had shown that nine was a sufficient number for most purposes. If one could finally manage to raise the number of categories to ten or twelve, then almost all interested parties would be satisfied. Producing a land utilization chart for an area the size of Iowa or Illinois would cost $80,000, compared with $200,000 using high-altitude reconnaissance planes, or $1,000,000 using mid altitude planes. This also provided the opportunity to draw up economically feasible land utilization charts of remoter areas.

It turned out that under optimal conditions their expectations were even exceeded to some extent. Occasionally they were able to identify fourteen to eighteen categories, but these were the exception. Furthermore, such a sharp differentiation was not always desirable. It was often sufficient to separate residential from industrial and agricultural areas. A work group at Dartmouth College, for example, got by with differentiating only eight types of land use.

The group analyzed the surface of Rhode Island which appeared very clearly on a picture of southern New England. Within forty hours they had charted an area of 1,200 square miles. This is equal to a work speed of 29 square miles per man per hour, quite an accomplishment for land evaluation. Most of the time was taken up with coloring the final map.

It is astonishing what could be recognized on the picture. Villages with a population of at least 7,000 could be identified. Individual factories and government buildings could be picked out. Bridges and canals, four-lane highways, railroad viaducts, and even golf courses appeared plainly. This richness of detail promises broad application of photography from outer space in city planning.

Only the future can show us how extensive the possibilities are in other places where the atmosphere is not as clear as in the broad expanses of the United States where the EROS testing grounds are situated. Only the future will show us whether the observation techniques can be improved to the point where for reasonable expense even smaller villages can be surveyed. However, the results already attained are exciting enough.

One can hardly believe what has already been achieved. Based on certain characteristics, scientists have been able to differentiate on the satellite pictures, industrial and residential sections on the thickly settled east coast of the United States. In some cases, they could even determine whether these residential areas were primarily settlements with single-family housing, or with multi-family housing.

One case which the authorities must have welcomed makes it obvious that exploration from outer space can have practical significance. Using various photographs from space they surveyed several cities in Arizona closely. Certain things became clear which had not been suspected at all until then. In Phoenix, for example, large areas were surveyed with a precision never before attained in this area. The results showed that some of the areas were much larger than recorded in official registers. The Office of Taxation immediately used this knowledge to send new tax bills to the property owners involved.

A comparison of pictures taken by Apollo 9 astronauts and the ERTS satellites was not less valuable. Because of the time elapsed between these two series of photographs, comparing them allowed technologists to make statements about how fast individual cities were growing and encroaching on valuable agricultural land. Within three years, the city of Tempe had expanded by 13.5 square miles. In the same period the expansion of Phoenix reached 12.3 square miles, and of Mesa, 7.7 square miles. The Americans are very concerned about the tendency of cities to spread at the expense of arable land. Perhaps further satellite picture evaluation will give us the means to develop an acceptable city planning for the future.

Until now only a few of the necessary conditions for this prevail; it is very difficult to draw up plans for settlement of a large area that have not already became dated. The above example of cities in Arizona proves this fact. Naturally the same results can be obtained by airplanes, but they are expensive. Nonetheless, such an experiment was once carried out on a large scale. In 1970, at the time of the last census, twenty six cities in the United States and Puerto Rico were photographed from an altitude of nine miles. This was repeated in 1972, and we can now compare the multispectral photographs. This will show whether airplane photography provides so many more valuable details for surveying cities that satellites will be unecessary, despite the fact that they are cheaper.

We should remember that satellites have already proven their value, if only in a few cases. Road maps of Tucson, Arizona were made using Gemini 5 pictures. Gemini 5 and 7, as well as Skylab pictures, registered the changes in the agriculture on Cape Canaveral, and so forth. The results often proved to be useful enough for practical application.

The Battle Against Air Pollution

Urban problems are different than they used to be. It is true that even at the time of the Romans there were cities with populations in the millions, but these were very rare exceptions, and could not have had the effects that we can see today. Even an extreme expansion of these settlements could not lead to a notable decrease in agricultural area, because there simply were not enough people. Today this has changed; the population has increased so much that the allotment of land must be carefully planned.

We must also consider another factor. In highly civilized countries, extensive industrialization accompanies the spread of cities. To maintain a high standard of living, goods must be produced. New factories shoot up over night, producing goods which are sold and used up. The final result is that the world is being poisoned and polluted. In many cases this has reached the limits of what is tolerable. Poisonous gases, dust, and ashes enter the atmosphere in large amounts. The consequences are often unforeseeable.

In the Fall of 1971, an extremely interesting discussion took place at a convention in Brussels. Space flight experts from many countries, among others the United States and Russia, came to grips with pollution. In the discussion, the Americans asked their Soviet colleagues how they could talk about observing the atmosphere from outer space. Whenever the Russians launched an observation satellite, the rocket gave off so much polluting exhaust that the satellite could probably only take measurements of that.

When we use normal rocket fuel we produce carbon monoxide and carbon dioxide exhausts. The first of these is a deadly poison, while the second causes the atmospheric temperature to increase, and thus affects the earth's climate. The Americans have partially switched to a combination of fuels (liquid hydrogen/liquid oxygen) which produces mainly water vapor, and is therefore better for the environment. Incidentally, this change was prompted by other considerations than keeping the atmosphere clean. To return to the convention once more: the Soviets countered by pointing out the noxious fumes from cars which poison American cities. In the Soviet Union this problem is still virtually unknown.

We really should take heed of the consequences, and leave our car in the garage, but no one wants to do that. Man's four-wheeled friends conspire with the many factories and chimneys to pour filth into the air; they are actually his worst enemy. For this reason, a smog warning service has had to be set up in many cities. When certain levels of air pollution are reached, this service can declare a state of emergency.

Is this the beginning of a problem that could lead to mankind's decline? Is it really the eleventh hour? Probably not. With a bit of optimism and a willingness on the part of responsible offices to take appropriate steps, our "air environment" can be restored to good condition. It is necessary however, that the sources of air pollution be located and eliminated. Hard times will then begin for the "environmental gangsters", and satellites can contribute to this.

In fact, they have already done so. The Americans have made a start. They looked at the pictures ERTS 1 took of an area in Virginia. These photos were so clouded that only small parts of the actual surface were visible. They then counted the smokestacks which were identified by clearly recognizable plumes of smoke. Their count reached approximately 10,000, which surprised them a bit, since the environmental offices had not registered so many smokestacks. Once again satellite pictures had proven their capabilities. Everybody who had broken the pollution laws must now count on paying a fine.

The pollution in Virginia is no worse than that in many other regions. Whether one looks at London or the Ruhr Valley, the degree of haziness on the pictures is always shocking. Experts think that the most valuable contribution of future earth reconnaissance satellites in the Federal Republic of Germany will be to register air pollution. Instead of data from a few selected points, we would get overviews of the entire polluted area. We would then be able to investigate and vigorously fight the pollution.

Our present technology alone is not adequate. ERTS 1 only sends a picture of a specific area of the earth every eighteen days, whereas we would need pictures at least every two or three days. We must also remember that in central Europe many photographs show only clouds, so that the actual rate of photographing would have to be even greater here. In the first year after the launch of ERTS 1 the Ruhr areas was only photographed twice from outer space! Perhaps an operational ERTS reconnaissance system will provide the means to solve this problem.

It is well known that many gases which reach the atmosphere are extremely poisonous, and thus dangerous. Carbon monoxide is the best example. The damage caused by this can easily be determined. However, the actual effects of other materials that pollute our environment, but are not directly poisonous, are completely unknown. Carbon dioxide is one of these. It causes what is called the greenhouse effect in the atmosphere: an increase in temperature and, thus, a gradual change in the climate. Its extent is still a source of much controversy.

Among others, the father of the hydrogen bomb, scientist Edward Teller, sees carbon dioxide as dangerous to the earth. He said that the carbon dioxide content in the atmosphere has increased by 2% since the advent of industrialization. An increase of 10% would cause the polar ice caps to melt, making the level of the seas rise. There are estimates which say that by the year 2000 many coastal cities and regions will be flooded, because the sea level will have risen five feet.

This thought would really be frightening, if other calculations did not show that an even larger increase in the carbon dioxide content of the atmosphere will cause only a negligible increase in temperature on earth. Which theory is correct? Nobody knows, since, fortunately, we have not yet reached this stage. It seems fairly sure that a temperature increase in the atmosphere is checked by a number of very small particles which help radiate excess warmth back into outer space. These particles, also a result of air pollution, exist in greater numbers today than ever before. Thus we must remember that man is presently helping change the earth's climate, but we do not know how it is changing. Perhaps in a few years satellites will provide the answer.

And Now the Forests. . .

We are becoming increasingly aware that we must think about the future. We cannot settle just anywhere on the land and senselessly destroy it. We must think of our children. If the problems of the world are so great today, they will be even worse tomorrow. What kind of inheritence can we leave? For the first time in the history of mankind these considerations are coming to the surface. It never occurred to the Egyptians or the Romans, or any other cultured peoples, that they could harm their descendants by reckless management of the environment.

Who ever worried when entire forests were chopped down? After all, there were enough trees in the world, and what could be the consequences of mass deforestation? It was ridiculous to talk about it! Today we see the results. Where lush vegetation grew thousands of years ago, there is now wasteland. These mistakes must not be repeated. This would be truly catastrophic. The forest areas of the earth must also be included in global management.

Our task is not easy. Forty five percent of all forests are in the tropics. Like the sub-polar regions in Canada and the Soviet Union, they are far too expansive to be surveyed. How can we sensibly determine which parts of these forests should be freed for felling without despoilation of the land? How can we prevent uncontrolled forest fires from burning up thousands of square miles?

Although these questions have not entirely been cleared up one thing is sure: we first need a map of the forests, a world forest atlas. Scientists have been waiting decades for such an atlas to appear, but the effort involved would be enormous. Many of the earth's forest areas have never been flown over by airplanes, let alone entered on foot. Consequently, the scientists' wish could not be fulfilled. The Freidburger Forestwirte project (Freiburg Forestry Project), aimed at compiling a worldwide forest atlas, was started in 1930. Today, more than forty years later, it is still incomplete.

This presents a real opportunity for earth reconnaissance satellites. With their television eyes, they penetrate into the outermost corners of the earth, they photograph the most remote regions. Since ERTS 1 was put into orbit, realization of the worldwide forest atlas has come closer. Within one year it could become fact. The technological earth reconnaissance satellite has shown that it is suited for this purpose. In evaluating its pictures, Canadian scientists have already been able to differentiate clearly the different types of forests. They have succeeded in identifying spruce, poplar, and willow trees, shavegrass, and uncultivated tundra. This is an important step.

The Soviet Union, like Canada, is interested in forest supervision from outer space. The Soviets cannot complain of a lack of forests, but they are afraid that uncontrolled cutting will lead to the same kind of devastation as has often happened elsewhere in the past. It is not surprising that at the same time they show an unusually strong interest in the earth's wastelands?

We know much too little about Soviet earth reconnaissance from outer space. They have carried out extensive surveys, primarily during their Soyuz flights. The cosmonauts' observations were directed at the most varied phenomena on earth, and, like the sightings of the American astronauts, were supplemented by ground crews and aircraft. An example is the observation of wastelands made by the crews of Soyuz 6, 7, and 8.

At the same time other satellites and space probes were directed at the earth. Zond 5 relayed a series of pictures back to the earth primarily showing the African continent. In spite of their low resolution, these pictures did provide valuable information. The scientists at Leningrad University produced a series of special maps of Africa. Among them was a geobotanical map.

The Soviets later turned mainly to geological investigations. Not only the Americans want to find new sources of raw materials. During the flight of Soyuz 9, the Russians examined test areas in the northern Caucasus, in the Caspian Sea, and on the Aral Sea, in Kazakhstan, and in western Siberia. At a convention in Constance, West Germany in 1970 the cosmonaut Sevastianov, who was one of the crew on this flight, reported on some of the earth reconnaissance experiments:

"We took valuable photographs and photometric readings which made a geological survey possible. The length of the individual routes reaches 600–1300 miles. Besides the serial pictures, we photographed individual areas, namely the Mangyshlak peninsula, the northern Caucasus, an area north of Lake Baikal, and others. These pictures from space provided information about regional formations. They showed regularities in the geological structures that influence the distribution of natural gas, oil, and other deposits, and which make possible further study of these structures."

Even if this statement, as is typical of Soviet space exploration, does not say anything, nonetheless, we can see clearly how the interests of the East and the West overlap.

For this reason the Americans and the Soviets signed an agreement in 1971 providing for the exchange of all information in the area of earth reconnaissance. The superpowers are moving together. Once again the realization that duplication of effort is not always necessary gained the upper hnd. This may be the beginning of a global study of the earth which will transcend political boundaries. Perhaps. Perhaps it may also extablish a realistic basis for a world forest atlas.

One event that happened several years ago will show why such an atlas is necessary. In its program for developing countries, West Germany was supposed to help Afghanistan build a power plant. At the same time, they were to erect a saw mill. However, during the planning stage it turned out that nobody knew where large forest areas existed in that country. There was no other way to clear up this question, but to first conduct a major air reconnaissance program. The costs were considerable, and the results showed that in the entire area under question, there were no large forests! This knowledge was bought dearly.

Even when we know where the forests are, we have not eliminated all the problems. Forests must also be supervised. We cannot allow uncontrolled forest fires to destroy our earth's precious trees. This problem has long been familiar in the United States and Canada where they have had air supervision for years. Using equipment sensitive to infrared light, they try to discover even the smallest fires at an early stage, so that the procedures necessary to extinguish them can be started.

The undertaking is not only tiresome and spotty, but also very uneconomical. ERTS pictures were therefore examined to see whether they could help in the early detection of forest fires. Many pictures clearly showed large trails of smoke which betrayed a raging fire. In one case, the forest fire, as it appeared on an ERTS picture, was compared with the results of conventional reconnaissance. The differences with respect to the extent of the fire reached 22%. This is definitely too much. It was not the ERTS pictures that were wrong, but rather the conventional measurements!

Even more important than discovering forest fires, is preventing them. Although this is only possible in the rarest of cases, we can know ahead of time that a certain area is threatened by forest fire. We must then be especially attentive. One such case where an endangered area was spotted from outer space occurred at the end of 1972.

When checking over an ERTS picture of the Oakland-Berkeley area, the scientists noted that a 3,000 acre eucalyptus grove was completely dead. The trees had fallen victim to a nine-day frost. In the following spring, 1973, an unusually large amount of rain fell on the same area. These two things together had consequences. The grass grew more lush than usual, and, when summer came, it dried out. Along with the dead eucalyptus trees, the grass constituted a tremendous fire hazard. The authorities had to institute protective measures, and made $4 million available for forest fire prevention.

The technological reconnaissance satellite, ERTS 1, has been surveying the earth since July 23, 1972.

13. All the Water in the World

No One Knows the Seas

"Next day we were sailing in slack winds through an ocean where the clear water on the surface was full of drifting black lumps of asphalt, seemingly never-ending. Three days later we awoke to find the sea about us so filthy that we could not put our toothbrushes in it. . . The Atlantic was no longer blue but gray-green and opaque, covered with clots of oil ranging from pin-head size to the dimensions of the average sandwich. Plastic bottles floated among the waste. We might have been in a squalid city port."

This report is from the well-known scientist Thor Heyerdahl and his crew who undertook to venture across the Atlantic from east to west on a papyrus raft. The report may sound unbelievable but it is true. These signs of "civilization" can be seen in the middle of the ocean far from all shipping lanes. The water there is like mud puddles. Man is on the verge of making a great mistake. He is killing his future, because he cannot live without the sea.

The oceans covering 140 million square miles, comprise about 71% of the earth's surface. Their influence on the world's climate is undisputed. Evaporation of water, the warm and cold ocean currents, and many other phenomena contribute to this end. The sea cleans the air, and contains a large portion of the food reserves which we will need sooner or later.

Investigating the ocean is very difficult precisely because of its expanse. There are surveying buoys and anchored surveying ships. Ships outfitted with the most modern research equipment criss-cross the seas. Yet all of this is only a drop in the bucket. Many secrets had to go unexplained, since a continuous wide-scale observation of the oceans was not possible with conventional means.

The launch of the weather satellite Nimbus 1 was the first turning point. Suddenly there was a means for observation. Everybody was aware of the limitations that still existed, but Nimbus 1, with its infrared equipment, was at least capable of regularly providing heat pictures of the mot remote areas of the seas, pictures on which we can clearly see the course of warm and cold currents. This was a beginning.

Charting these ocean currents had always been a problem for the scientists. Just measuring them was tedious and expensive enough, but what made it worse was that they were continually changing their course. Thus the research ships had to go out repeatedly to take measurements. The currents were simply too important (especially since they determine the climate in many parts of the world) to dispense with information about them. Central Europe, for example, owes its mild climate mainly to the Gulf Stream.

It is obvious that much that occurs in the oceans' expanses remained hidden from Nimbus 1. The resolution strength of its equipment was much too limited. Furthermore, many details can only be observed in certain wave lengths, and these were poorly covered by Nimbus. In spite of that, oceanographers did not initially expect very much from the detailed photography that was carried out in manned space flights. After all, what could one expect?

The scientists were wrong. They could not have been more surprised when they saw the pictures taken by the Gemini and Apollo astronauts. There was so much that they could identify! Suddenly, a new world seemed to be opening up before them. For example, some pictures showed quite plainly how the Gulf Stream picked up an extensive area of polluted water near Cape Hatteras, North Carolina and carried it along to Europe. They had not expected they would be able to distinguish such details. This showed them entirely new prospects. Indeed, research into the oceans seems to have taken a great leap forward with the introduction of the technological earth reconnaissance satellite, ERTS 1.

Only time will reveal how incorrectly one has evaluated what goes on in the oceans' expanses. We had felt that we could pour our garbage into the ocean without a second thought. The sea would quickly take care to spread out this garbage, so that it would disappear. There are such approved garbage dumps everywhere in the world. One of them is along the coast of New York, another is in the North Sea near Helgoland.

The scientists were rather shocked when they looked at an ERTS photograph and discovered an enormous garbage dump which was not supposed to exist. A ship had just dumped a load of excess acid into the ocean near New York. The acid should have dissolved in the sea water, so that it would no longer be discernable. This had not happened. The acid remained where it was for quite a while, creating a health hazard for the coastal population. Nobody had reckoned with this.

Looked at in this light, Thor Heyerdahl's description becomes interesting again. If the ocean is such a poor disperser of garbage, then we can understand why the seas are so filthy far from shipping routes. We should be careful to remember this. We cannot soil the seas and go unpunished. There are already enough problems on the coasts. For example, in Sweden the oil along the coastal areas has spread to such an extent that the country saw no other way out than to supervise the damage continually. They employed the German firm Dornier which uses the most moern methods of radar for this purpose.

In other parts of the world, the coasts are not quite as threatened as in Sweden, but this can easily change. There is no lack of filth. Many places no longer have even a trace of clean water, especially where the rivers flow into the ocean, where man's entire refuse is poured into the open sea. The pollution at the mouth of the Elbe is depressingly clear on an ERTS picture. We can see similar things over and over again on pictures from outer space.

This seems to be an appropriate place to put satellites to work, supervising coastal regions. Pictures from outer space clearly show how far large rivers reach into the ocean, because the coastal waters are so shallow that the entire structure of the ocean bottom can be plainly seen at a glance. Gemini 4 and 5 pictures were even used to survey the geography of the Carribean. This would have taken weeks with conventional methods, since the ocean floor is inaccessible to the earthbound observer, and can only be explored at single locations every so often.

Particularly impressive are the changes in underwater geography that can be observed in coastal regions. High and low tides, rivers that pour their water into the sea, and the various ocean currents contribute constantly to agitate the ocean floor. Hydrologists do their best to map all of these changes as quickly as possible, because ship captains must know exactly where the water is deep enough to safely navigate their ships. Charting takes too long. Usually ships' charts are already out of date when published.

It is not surprising that hydrologists have finally gotten valuable aid from the ERTS photographs. A test case was made in Schleswig-Holstein, West Germany, where an ERTS photograph was compared with the latest coastal charts. Many differences showed up. Among other things, the ERTS picture revealed several sand banks which had not yet been recorded. Introduction of operational earth reconnaissance satellites will make coastal shipping even safer than it is today.

The fishing industry is also interested in pictures of coastal regions. Many rich fishing areas are located there, and these are often associated with characteristic structures on the ocean floor. If one knows where to find these structures, then one also knows where to fish. A Gemini photograph showing southern Texas and southwestern Louisiana is an example.

On this picture we can plainly see currents carrying sediment-laden water. During certain seasons shrimp prefer these sediment deposits as playgrounds, and fishing here is productive.

We must of course know something about the habits of schools of fish in order to make good use of these pictures from outer space. Sediment deposits do not always indicate a good fishing area. As an example of this we can cite an ERTS 1 picture taken in July 1972 which also showed unusually high sediment deposits on the New England coast. A few weeks after this picture was shot the State of Massachusetts prohibited shellfishing in precisely this area, since several dozen people had come down with shellfish poisoning. The loss caused by prohibition of shellfishing ran to about $1 million per week. What had happened?

Apparently, the build up of sediment deposits had attracted a special type of poisonous plankton to this area. The shellfish were infected by the plankton, and people were then poisoned when they ate the shellfish. Thus the sediment deposits had proven to be unhealthy. Because of this incident, scientists are presently studying the ERTS pictures more closely to see if they can discern traces of plankton. This would then provide a warning system against such occurrences.

"Fishing from outer space" has often been discussed, since according to some estimates approximately 80% of all animal life lives in the sea. The ocean will be our source of food for the future. If we succeed now in finding fishing grounds by satellite, and thus in decreasing the time fishing ships must search, it would be a tremendous gain. Twenty to thirty percent of the time ships are on the open seas is spent looking for new fishing grounds.

There are a number of ways we can find fish by using satellites. We can ignore the example of sediment deposits since, outside the coastal areas, the sea is so deep that not even satellites can show the structure of the ocean floor. However, there are other possibilities that promise success. As a matter of fact, one of these has just been applied. It is based not on locating the fish, but rather on finding areas rich in sea food.

The idea is very simple. Fish must lie, and what do they live on? They live on plankton, for one thing. Plankton changes the color of the water. It tints the ocean green, and this tinting can often be seen very plainly on ERTS photographs, as for example in one photograph of the Gulf of Mexico. The exact coordinates of the tinted area were determined by using this photograph, and were radioed to a research ship cruising in the area. The ship steamed to the indicated spot and threw out its fishing nets. The catch was unusually large.

Other possibilities for "fishing from outer space" have until now only been discussed theoretically, even though they are also very promising. For example there is the possibility of discovering sources of heat in the sea using infrared sensors. Large schools of fish often appear near them. Or there is the method of observing luminescence, that is the emission of light by certain fish, by using residual light photography which can magnify light 55,000 times. The light given off by larger schools would be sufficient to spot them from outer space.

Maybe we will even succeed someday in discovering schools of fish by studying the surface structure of the sea, even though this possibility is very vague. Such obervation of the sea's surface is still in its infancy. It was first accomplished during the Skylab flights. The astronauts scanned the ocean's surface with an active radar set, and the height of the waves could be determined from the reflected signals. They got some excellent results. The purpose of the experiment was actually to get information about wind speed on the high seas, since wind speed and wave height are closely connected. Thus they could obtain valuable data important not only for meteorology, but also useful for shipping. Selecting routes which lie favorably with respect to wind could make ships' journeys shorter and cheaper.

Our Most Valuable Possession

The seas contain 97% of the world's water, a material which is absolutely essential to maintain life. Although it is present in abundance, water is still the earth's most valuable possession. Sea water is salty, however, and has proven to be unusable for most purposes. So far all attempts to desalinate sea water economically have failed.

Nonetheless the situation would not seem bad at all, if at least the remaining 3% of our water were freely available. Unfortunately, this is not the case. Ninety five percent of this three percent is solidified as ice, mainly in the polar ice caps. What is actually left for consumption is a very small percentage, and this is not distributed evenly over the surface of the earth.

Agricultural land must be irrigated with this small amount of water. It also provides electricity, and is used for the personal needs of men, animals, and plants. Our needs are not exactly small. They reach tremendous proportions, especially in industrialized countries, because of the methods of production. In the United States, water use averages six tons per inhabitant per day! This means that the water content of the earth must be managed well, not only in industrial countries, but even more so in the hot regions of the world where mismanagement of the water supply can easily lead to catastrophe.

In the last few years, we have made many attempts to deal with this problem. New sources of fresh water have been found, new methods have been invented to control the flow of water so that its distribution could be optimized. Precautionary measures have been taken not to lose too much of this precious possession through carelessness. Although we have had some successes, we have also found that conventional methods are sharply limited. In many cases, satellites have been able to fill the existing gaps.

The supervision of extensive lake areas is one example. How can one keep all the small rivers and lakes in Minnesota under supervision? Only comparatively few people live in this region, where there are more than ten thousand lakes! Keeping them under continual supervision with satellites has proven quite possible. Evaluating ERTS pictures of Canada showed that we could determine whether water was clean or polluted – even in lakes with a diameter of only 400 feet.

Another example: in one region of South America with a surface area of 170,000 square miles, the scientists discovered thirty six new lakes while evaluating thirty one ERTS pictures. In addition, hundreds of moist zones were found, which promised fertile agricultural areas for the future. New water means new possibilities for life. Several areas on the earth are dependent upon such isolated finds.

In many countries such discoveries are still possible. Even ground water can be spotted on satellite pictures. Along the Senegal River we can see numerous small dark strips jutting out into the dry desert. Drilling for ground water in these areas promises some success. This is one of those regions where the inhabitants depend upon ground water.

The future will definitely bring many new possibilities for the use of earth reconnaissance satellites. A 1965 flight over Hawaii's volcanoes proved this. Nobody was thinking about ground water at that time, so that the technologists were very surprised when thermal infrared measurements showed sudden changes in the temperature of the sea along the coast. In many places the temperature suddenly sank by as much as 20° F. When they examined this phenomenon, they discovered that fresh cold water was pouring into the sea. This discovery was a boon for Hawaii which needs fresh water to exist, like all small islands.

Unfortunately, for various reasons a good deal of our fresh water is not available to mankind. We cannot even blame anyone for this. What happened in the dry desert areas of Arizona is a good example. Satellite pictures showed that some of the water flowing through the irrigation ditches never reached its destination, but trickled unused into the sand. Nobody had realized that before. On the spot examination revealed that the irrigation canals were not 100% watertight, but were filled with numerous small holes.

Even more shocking is a report about the fate of the Mekong Valley. Since the Americans bombed the area for years, and devastated the forests to destroy the enemy's cover, the water ecology of the country is ruined. The artificially produced clearings cause large amounts of water to evaporate. Can this mistake be corrected? No one knows. Furthermore, no one ever would have discovered anything at all about the consequences of bombing ten million acres of forest area, if ERTS 1 had not revealed this.

Nothing escapes the satellite. It even saw the pollution in Lake Champlain, on the border between New York and Vermont, from about 560 miles altitude. This picture has made history. Vermont has decided to add it to the evidence it is using in a civil suit against New York State.

On the New York shore of the lake, directly north of Fort Ticonderoga, the International Paper Corporation has a plant that pumps its waste water directly into the lake. Vermont wants to stop them from doing that, because its shores have also been affected. It does not want Lake Champlain to suffer the same fate that Lake Erie has experienced. This lake, on the border between Canada and the United States, was unable to absorb 2.75 million tons of garbage annually, and is now biologically dead.

While it is true that this ERTS picture is only a small piece of evidence in Vermont's case, there is plenty of other, and better evidence, enough to condemn New York. Perhaps the importance of the picture lies in the fact that for the first time, a satellite photograph is being used as evidence in a public trial. This could set a precedent.

Satellites, especially ERTS 1, have shown in many cases that they can give valuable help in supervising the water content of our earth. Their greatest contribution might be the prevention of catastrophes due to floods caused by snow suddenly melting in the Spring. When scientists started taking measurements of snow boundaries using conventional methods, they were able to realize an annual profit of $1 million for a single northwestern city in the United States by using the water from melting snow to produce electric energy. How great will the profits be when measurements of snow fields are based on ERTS photographs?

It has been known for years that satellites can be used productively to this end. Even pictures from the weather satellites with their poor resolution were able to show the snow boundaries so exactly that there was an error of only 5%. This gives us a good estimate of the amount of snow in a certain area, and allows us to draw conclusions on the amount of water that will result when the snow melts during the Spring thaw.

During the course of Apollo 9's flight, they were even able to decrease the percentage of error. Among other things, the astronauts on this mission had the task of determining the quantity of snow that had fallen in the Sierra Nevadas and in California's mountains. They succeeded beyond all expectations, making it possible to institute preparations in time for the thaw, and to prevent damage by widespread flooding. In the future ERTS 1 is supposed to assume this task.

We are of course only dealing with experiments, but we do expect to gain greater insight from them. The first evaluations of the pictures have shown that the snow boundaries could be determined with a precision of about 200 feet. This is tremendous. If estimates of the snow volume based on photographs are done at all carefully, then the time has come when we should establish a continuous flood warning service. This will be an improvement over the conventional type service, because it can work faster and can include areas which are not accessible to ground observation. In the United States this means every area where all kinds of surveying equipment are forbidden by the Wilderness Act.

The Dangerous World of Ice

Not far from the coast of Alaska, the Bering Glacier

stands out impressively from its surroundings. If we ignore a dam on the west side behind which a lake has been formed, there is nothing special about it. If one looks at the glacier more closely, however, we suddenly see that it is one of those that remains quiet and immobile for a long time, only to start moving again unexpectedly and at a great speed. A velocity of one yard per hour is not uncommon.

That is the unusual thing about the Bering Glacier: one day it will start moving again, taking everyone by surprise. Then the dam will break, and the floods from the lake will rush into the valley burying everything that lies in their path. As of yet there is no way to give advance warning.

However beautiful the ice is, its dangers are many. Usually people do not pay much attention to them, because the largest ice fields are located in the uninhabited areas of the world. (Ninety per cent of the 7,200,000 cubic miles of ice fields cover Antarctica. Nine per cent lie in Greenland, while only the remaining one per cent is distributed over the rest of the Arctic and on high mountains.) However, when ice strikes, it strikes with a vengeance.

Today we have almost forgotten the legendary collision of the enormous passenger ship, the Titanic, with an iceberg on April 15, 1912. At that time the whole world was in an uproar. Fifteen hundred people lost their lives when the ship went down with everyone aboard, and there was no chance for rescue. All help came too late.

The event is forgotten mainly because no similar collision has happened since then in the north Atlantic, where the U.S. Coast Guard keeps watch. The sinking of the Titanic gave the impetus to form an International Ice Patrol in 1914. This patrol still exists today, financed by seventeen countries whose ships cross the north Atlantic. The U.S. Coast Guard uses its ships and airplanes to keep an eye on the ice conditions that are predominant there. Twice daily it radioes the results, warning ships throughout the world.

For some time now several nations have gradually become more interested in proceeding further into the north polar regions with their ships. Here we do not mean research ships, bur normal merchant vessels. The Soviets especially are interested in developing the northwest passage into a secure shipping lane leading from Europe, past Siberia, to Japan. The Americans are mainly interested in opening up the northwest passage so that they can ship the oil they drill in Alaska to the refineries on the east coast. They undertook this experiment in October, 1969.

The development of the northern polar region is thus proceeding apace. However, there is no U.S. Coast Guard there to keep track of the icebergs and driving ice floes, and warn the ships. In this area, there is only one solution: the ice conditions must be monitored by satellite.

In an earlier chapter we talked about a trip by a Soviet research ship through the Antarctic, which was competently directed by Soviet weather satellite pictures. This example shows that today there are already ways to keep polar ice conditions under supervision. Weather satellite pictures, however, are too coarse to show small icebergs. Merchant shipping in the northern polar region must be made safe. For this reason we will have to refine our observations sooner or later.

Now there is ERTS 1. The pictures that this satellite has taken of various ice regions are fantastic. They reveal many interesting details which escape weather satellites. The location of glaciers, especially in the north polar region, can now be researched.

As in many other cases, one could object that at present earth reconnaissance satellites only fly over any given area every eighteen days, so that continuous supervision of ice on the northern sea passages is not feasible. This objection is not entirely valid. To be precise, it is only true for the equatorial regions. The closer we approach the poles, the more the photographs overlap, so that we have a higher frequency of photography. In practice, the extreme polar regions are photographed during every orbit of the satellite!

We must bear in mind: the optical equipment of ERTS 1

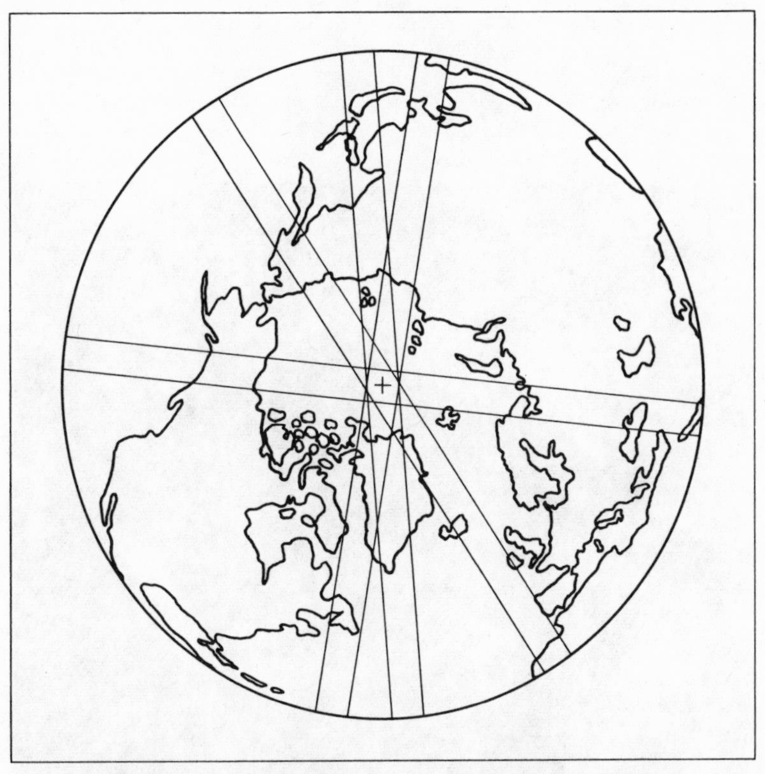

A look at the North Pole clearly demonstrates that even if a satellite at the equator only photographs a specific area every eighteen days, it can shoot regions near the poles significantly more often, because of the overlapping of individual "flight lanes". (Schematic diagram based on an orbital inclination of 90°.)

is well-suited for supervising ice, but is not nearly as good as it could be. Research in the last few years has shown that microwave sensors are especially appropriate. These are sensors that make their measurements in an extremely short wave length range of the electromagnetic spectrum. The Americans and the Soviets are working together to find the best sensors for use in outer space. For this purpose they

Acid residue in front of New York Harbor! This formation, seen from outer space, shows how long lasting the effects of chemical wastes are when dumped in the ocean.

together undertook a large-scale experiment in the Bering Sea from February 15 to March 7, 1973. It was intended to test new microwave sensors.

In spite of everything, even equipment that is not the best can often be very useful. The U.S. Coast Guard has been employing weather satellite pictures to help prepare their ice reconnaissance flights for a long time. They can tell from these pictures which areas of the North Atlantic need special reconnaissance flights. When their airplanes have a definite goal, the entire task of ice supervision is more productive.

Since the introduction of the first Tiros satellites we have known that reconnaissance of large ice fields using weather satellites has advantages not possessed by airplane reconnaissance, despite the poor power of resolution in the pictures. Even at the time of Tiros 2, scientists tested the possibility of distinguishing ice fields from the clouds which lay over them. During the flight of Tiros 4 they undertook a project in which airplane photographs were made at the same time as the satellite pictures, and conventional ice maps were drawn up. The project bore the name TIREC (Tiros Ice Reconnaissance). The results confirmed the importance of satellite pictures to such an extent that the experiments were continued. By April 28, 1963 the first satellite observations were incorporated directly into the official report of ice conditions.

In the long run, earth reconnaissance satellites will make ice reconnaissance even more effective. This is yet another field of application. Nobody can say for sure that it will be the last one. Earth reconnaissance from outer space is presently at a turning point. We have already looked at numerous possibilities. However, it is still to early to deliver the final verdict. When the first communications satellite was launched, who ever suspected that satellite communications would quickly expand and become irreplaceable? When the first Tiros satellites were launched who could guess the possible application for satellite meteorology? We should look to the future of space exploration with anticipation. Undoubtedly it still holds many surprises in store for us.

INDEX